AF559774

TEXTBOOK OF INDUSTRIAL MANAGEMENT

TEXTBOOK OF INDUSTRIAL MANAGEMENT

By

Shashi Kant Yadav

DISCOVERY PUBLISHING HOUSE PVT. LTD.
NEW DELHI-110 002

Published by:
Tilak Wasan

DISCOVERY PUBLISHING HOUSE PVT. LTD.
4831/24, Ansari Road, Prahlad Street
Darya Ganj, New Delhi-110002 (India)
Phone: +91-11-23279245, 43764432
Fax: +91-11-23253475
E-mail: parul.wasan@gmail.com
info@discoverypublishinggroup.com
discoverypublishinghouse@gmail.com
web: www.discoverypublishinggroup.com

***First Edition:* 2011**
ISBN: 978-81-8356-842-5

Textbook of Industrial Management

Printed at:
Mehra Offset Press
Delhi

Preface

Industrial management is one of the most attractive specialization for mechanical engineering graduate. This course is also a core course in almost all branches of engineering. However at some places, its name differs a little bit, for example engineering management; engineering economy and management, industrial management etc. This book is aimed to cater to need of these courses.

The coverage of the topics are planned to integrate concepts, models, numerical examples and industrial applications. Sufficient care has been taken to present the recent developments and trends in each area. Despite this, it is always possible that some relevant topics outlined in the syllabus of few universities are still missing. We would appreciate if such gaps are brought to the attention of the author so that improvements may be undertaken in the next edition.

Author

Contents

1

Information Quantities and Resource Levelling

INTRODUCTION

Marginal and Joint Entropies : In previous sections we have discussed the measures of uncertainty for one-dimensional probability distribution only, *i.e.*, the amount of information obtained by the occurrence of the simple event. In this section, we shall extend the concept to two-dimensional probability distribution to help in the simultaneous study of both transmitter and receiver in a system.

Let two finite discrete sets $X = \{x_1, x_2, ... ,x_m\}$ and $Y = \{y_1, y_2, ... ,y_n\}$ exists, where x_i's and y_i's denote the massages transmitted and received respectively. Thus, if

(i) $x_i = y_j$, then the message received is correct, and

(ii) if $x_i \neq y_j$, then the message received is not correct due to any reason.

Let P (x_i) and P (y_j) be the probability of message x_i transmitted and message y_j received, respectively; P $\{x_i, y_j\}$ be probability of the joint event that message x_i is transmitted and message y_j is received. The values of these probabilities are calculated by calculating the average of the messages transmitted and received.

The probability P $\{x_i, y_j\}$ can be helpful in estimating the effect of disturbances during communication.

Thus when P $\{x_i, y_j\} = 1$ for $x_i = y_j$, the communication is considered without disturbance. The joint probability distribution on the set X, Y can be defined as

$$P(x_i, y_j) = P[x_i \in X \text{ occurs and } y_j \in Y \text{ occurs}]$$

In matrix notation, it can be expressed as

$$P(X, Y) = \begin{bmatrix} x_1y_1 & x_1y_2 & \cdots & x_1y_n \\ x_2y_1 & x_2y_2 & \cdots & x_2y_n \\ \vdots & \vdots & & \vdots \\ x_my_1 & x_my_2 & \cdots & x_my_n \end{bmatrix}$$

From this joint distribution, the marginal (individual) probabilities can also be determined as follows:

$$P\{x_i\} = P\{x_i \in X \text{ occurs}\} = \sum_{j=1}^{n} P\{x_i, y_j\},\ i = 1, 2, \ldots, m$$

$$\text{and } P\{y_j\} = P\{y_j \in Y \text{ occurs}\} = \sum_{i=1}^{m} P\{x_i, y_j\},\ j = 1, 2, \ldots, n$$

Now, it is also possible to determine the marginal entropy functions of X and Y as follow:

$$H(X) = -\sum_{i=1}^{m} P\{x_i\} \log P(x_i\}$$

$$H(Y) = -\sum_{j=1}^{n} P\{y_j\} \log P\{y_j\}$$

The entropy H (X) and H (Y) measures the uncertainty of the message transmitted and received irrespective of the massage received and transmitted, respectively.

The joint entropy function of X and Y which represents the entropy of the joint distribution of message transmitted and received is given by

$$H(X, Y) = -\sum_{i=1}^{m}\sum_{j=1}^{n} P\{x_i, y_j\} \log P\{x_i y_j\}$$

This function measures the uncertainty of the message transmitted and received simultaneously.

CONDITIONAL ENTROPIES

There may be situations where message received (y_j) occur in conjunction with any of the message transmitted (x_i) and *vice-versa*. In such cases the knowledge of conditional entropies H (X | Y) and H (Y | X) would be quite helpful.

Case I

Let $H(X \mid Y) = P\{X = x_i \mid Y = y_j\} = \dfrac{P\{X = x_i \cap Y = y_j\}}{P\{Y = y_j\}}$;

$$= \frac{P(x_i, y_j\}}{P\{y_j\}} \text{ for all } i = 1, 2, \ldots, m; \; j = 1, 2, \ldots, n \qquad \ldots(1)$$

denote the conditional probability of message transmitted xi when it is given that message received is, y_j. The amount of information provided by this simultaneous transmission of message and receipt of the message is given by

$$I(x_i, y_j) = -\log P\{x_i, y_j\}$$

Thus, the amount of information obtained with the transmission of the message x_i when message y_j has already been received is given by

$$I(x_i \mid y_j) = -\log P\{x_i \mid y_j\} = -\log \frac{P\{x_i, y_j\}}{P\{y_j\}}$$

Hence the average amount of information provided by the transmission of message x_i for any value of i (i = 1, 2, ..., m) when it is known that message y_j has already received is given by

$$H(X \mid y_j) = -\sum_{j=1}^{n} \frac{P\{x_i \cap y_j\}}{P\{y_j\}} \log \frac{P\{x_i \cap y_j\}}{P\{y_j\}}$$

$$= -\sum_{j=1}^{n} P\{x_i|y_j\} \log P\{x_i|y_j\} \qquad \ldots(2)$$

The average of the conditional entropy $H(X \mid y_i)$ for all values y_j, (j = 1, 2, ..., n) is then given by

$$H(X \mid Y) = -\sum_{j=1}^{n} P\{y_j\} H\{X|y_j\}$$

$$= -\sum_{j=1}^{n} P\{y_j\} \left[\sum_{i=1}^{n} P\{x_i|y\} \log P\{x_i|y_j\} \right]$$

$$= -\sum_{j=1}^{n} \sum_{i=1}^{m} P\{y_j\} P\{x_i|y_j\} \log P\{x_i|y_j\}$$

$$= -\sum_{j=1}^{n} \sum_{i=1}^{m} P\{x_i|y_j\} \log P\{x_i|y_j\} \qquad \ldots(3)$$

Case II : Similarly, the average of the conditional entropy H (Y | x_i) for all values of x_i, (i = 1, 2, ... ,m) is given by

$$H(Y \mid X) = -\sum_{i=1}^{m}\sum_{j=1}^{n} P\{x_i\}P\{y_j|x_i\}\log P\{y_j|x_i\}$$

$$= -\sum_{i=1}^{m}\sum_{j=1}^{n} P\{x_i|y_j\}\log P\{y_j|x_i\} \quad ...(4)$$

Theorem 1:

Establish the following results for two dimensional probability distributions:

(a) H (X, Y) = H (X) + H (Y) if and only if X and Y are independent.

(b) H (X, Y) = H (X | Y) + H (Y) = H (Y | X) + H (X).

(c) H (X) ≥ H (X | Y); H (Y) ≥ H (Y | X).

Proof:

(a) By the definition of entropy function H, we have

$$H(X) + H(Y) = -\sum_{i=1}^{m} P\{x_i\}\log P\{x_i\} - \sum_{j=1}^{n} P\{y_i\}\log P\{y_j\}$$

$$= -\sum_{i=1}^{m}\left\{\sum_{j=1}^{n} P\{x_i, y_j\}\log P\{x_i\}\right\} - \sum_{j=1}^{n}\left\{\sum_{i=1}^{m} P\{x_i, y_j\}\log P\{y_j\}\right\}$$

$$= -\sum_{i=1}^{m}\sum_{j=1}^{n} P\{x_i, y_j\}\log P\{x_i, y_j\}$$

Also $$H(X, Y) = -\sum_{i=1}^{m}\sum_{j=1}^{n} P\{x_i, y_j\}\log P\{x_i, y_j\}$$

Since $\left[\sum_{i=1}^{m} P\{x_i\}\right]\left[\sum_{j=1}^{n} P\{y_i\}\right] = \sum_{i=1}^{m}\sum_{j=1}^{n} P\{x_i, y_j\} = 1$, therefore it follows that H (X, Y) = H (X) + H (Y) holds if and only if above equality holds for i and j. This condition also indicates that X and Y are independent.

(b) The following relationship exists among marginal, joint and conditional probabilities:

$$P\{x_i, y_j\} = P\{x_i \mid y_j\}\, P\{y_j\} = P\{y_j \mid x_i\}\, P\{x_i\} \quad ...(5)$$

and $\log P\{x_i, y_k\} = \log P\{x_i \mid y_j\} \mid \log P\{y_j\}$

$$= \log P\{y_j \mid x_i\} \log P\{x_i\} \qquad ...(6)$$

Thus $$H(X, Y) = -\sum_{i=1}^{m}\sum_{j=1}^{n} P\{x_i, y_j\} \log P\{x_i, y_j\}$$

$$= -\sum_{i=1}^{m}\sum_{j=1}^{n} P\{x_i, y_j\} \log[P\{x_i|y_j\}P\{y_j\}]$$

$$= -\sum_{i=1}^{m}\sum_{j=1}^{n} P\{x_i, y_j\}[\log P\{x_i|y_j\} + \log P\{y_j\}]$$

$$= -\sum_{i=1}^{m}\sum_{j=1}^{n} P\{x_i, y_j\} \log P\{x_i|y_j\} - \sum_{i=1}^{m}\sum_{j=1}^{n} P\{x_i, y_j\} \log P\{x_i|y_j\}$$

$$= H(X \mid Y) + H(Y)$$

Similarly, it can also be proved that $H(Y) \geq H(Y \mid X)$.

EXPECTED MUTUAL INFORMATION

The mutual information of the message transmitted x_i and the message received y_j is defined as follows

$$h(x_i, y_j) = \log \frac{P\{x_j|x_i\}}{P\{y_j\}} = \log \left[\frac{P(x_i, y_j)}{P\{x_i\}P\{y_j\}}\right]; \; i = 1, 2, ..., m$$
$$j = 1, 2, ..., n$$

The following two cases may arise:

(i) If no information is provided by one event about the other, then $h(x_i, y_j) = 0$, for all i and j. In other words, if events X and Y are independent, then one event cannot give information about the other.

(ii) If $h(x_i, y_j) > 0$ or < 0, for a fixed x_i, then more or less information, respectively, is received from the transmission of the message x_i given that message y_j has already been received. This is different from the independence pattern of a set of messages transmitted $X=\{x_1, x_2, ..., x_m\}$ and set of messages received $Y = \{y_1, y_2, ... y_n\}$.

Now, the expected mutual information or the averages amount of information about X that is provided by the occurrence of Y event may be defined as:

$$I(X \leftarrow Y) = \sum_{i=1}^{m}\sum_{j=1}^{n} P\{x_i, y_j\} h\{x_i, y_j\}$$

$$= \sum_{i=1}^{m}\sum_{j=1}^{n} P\{x_i, y_j\} \log \frac{P\{x_i | y_j\}}{P\{x_i\}}$$

$$= \sum_{i=1}^{m}\sum_{j=1}^{n} P\{x_i, y_j\} \log \left[\frac{P\{y_j | x_i\} P\{x_i\}}{P\{y_j\} P\{x_i\}}\right]$$

$$= \sum_{i=1}^{m}\sum_{j=1}^{n} P\{x_i, y_j\} \log \left[\frac{P\{y_j | x_i\}}{P\{y_j\}}\right] = I(Y \leftarrow X)$$

This shows that the average amount of information that an event Y provides about the occurrence of an event X is equal to the average amount of information that an event X provides about the occurrence of an event Y.

Remarks:

1. The expected mutual information is always non-negative. Thus, on the average mutual information is not misleading, the knowledge of the knowledge of the occurrence of an event in one set will on the average provide information about the occurrence of an event in the other set.
2. Expected mutual information measures will be zero if and only if X and Y are independent.

Theorem 2:

(a) $I(X, Y) = H(X \mid Y) = H(Y) - H(Y \mid X)$

(b) $I(X, Y) = H(X) + H(Y) - H(X, Y)$.

Proof:

(a) We know that

$$H(X) - H(X \mid Y)$$

$$= \sum_{i=1}^{m}\sum_{j=1}^{n} P\{x_i, y_j\} \log P\{x_i\} + \sum_{i=1}^{m}\sum_{j=1}^{n} P\{x_i, y_j\} \log P\{x_i | y_j\}$$

$$= \sum_{i=1}^{m}\sum_{j=1}^{n} P\{x_i, y_j\} \log P \frac{P\{x_i | y_j\}}{P\{x_i\}} = I(X, Y).$$

Similarly, it can also be proved that I (X, Y) = H (Y) – H (Y | X)

(b) The proof for the equality

I (X, Y) = H (X) + H (Y) – H (X, Y) follows form Theorem 1.

Axiom of an Entropy Function

1. The entropy function takes its maximum value when all the events have equal probabilities. That is,

$$\text{Max } H(p_1, p_2, \ldots, p_n) = H(1/n, 1/n, \ldots, 1/n)$$

where $p_1 = p_2 = \ldots = p_n = 1/n$

2. The information provided by the joint occurrence of the pair (X, Y) is equal to the sum of the information provided by the occurrence of X and that provided by the occurrence of Y given that X has already occurred, that is

$$I(X, y) = H(X) + H(Y \mid X)$$

3. The entropy of the function remains unchanged with the addition of an impossible event in that function, that is

$$H(p_1, p_2, \ldots p_n, 0) = H(p_1, p_2, \ldots, p_n)$$

4. The entropy is continuous with respect to all its arguments.

PROPERTIES OF AN ENTROPY FUNCTION

The following are the four basic requirements to be satisfied by the logarithmic form of the entropy function $H(p_1, P_2, \ldots, p_n)$,

1. *Monotonically* : $H(1/n, 1/n, \ldots, 1/n) = f(n)$ is a monotonically increasing function of n,

i.e., $n_1 < n_2 \rightarrow f(n_1) < f(n_2);\ n_1, n_2 = 1, 2, \ldots$

2. *Additivity* : An event; say x_n with probability p_n can be sub-divided into a number of mutually exclusive events, $e_1, e_2, \ldots e_m$ each with probability $p(e_1), p(e_2), \ldots, p(e_m)$ such that

i.e., $p_n = \sum_{i=1}^{m} p(e_i)$

and $H(p_1, p_2, \ldots, p_{n-1}, e_1, e_2, \ldots, e_m)$

$$= H(p_1, p_2, \ldots, p_n) + p_n H\left(\frac{e_1}{p_n}, \frac{e_2}{p_n}, \ldots, \frac{e_m}{p_n}\right)$$

3. *Grouping* : H (p_1, p_2, ... ,p_n)

$$= H(p_1, p_2) + p_1 H\left(\frac{p_1}{p_1}, \frac{p_2}{p_1}, \ldots, \frac{p_r}{p_1}\right) + p_2 H\left(\frac{p_{r+1}}{p_2}, \frac{p_{r+2}}{p_2}, \ldots, \frac{p_n}{p_2}\right)$$

where $$p_1 = \sum_{i=1}^{r} p_i \text{ and } p_2 = \sum_{i=r+1}^{n} p_i$$

4. *Continuity* : H (p_1, p_2, ... ,p_n) is continuous function with respect to all p_i's. That is, any change in p_i's of events will not bring any change in the value of the function.

CHANNEL CAPACITY, EFFICIENCY AND REDUNDANCY

Channel Capacity

The amount of average mutual information processed by the channel in a communication system is defined by I (X, Y) = H (X | Y). As the information processed by a channel depends upon the input probability distribution P (x_i) of x_i's, therefore it can be varied until the maximum of I (X, Y) is reached. Hence channel, say C, can be defined as,

$$C = \text{Max } I(X, Y) = \text{Max } \{H(X) - H(X \mid y)\}$$

for all P $\{x_i\}$.

But for a noise free channel, we have

$$I(X, Y) = H(X) = H(Y) \text{ and } I(X, Y) = H(X, Y)$$

Therefore, channel capacity C in this case may be redefined as

$$C = \text{Max } I(X, Y) = \text{Max } [H(X)]$$

$$= \text{Max}\left[-\sum_{i=1}^{m} P\{x_i\} \log P\{x_i\}\right] = -\log\left(\frac{1}{n}\right)$$

$$= \log n \text{ bits per symbol (or second)}$$

This is due to the fact that maximum of H(X) occurs when

$$P(x_1) = P(x_2) = \ldots = P(x_n)$$

Special Types of Channels

(i) *Binary Symmetric Channel* : The channel capacity for the channel matrix

$$\mathbf{P} = \begin{bmatrix} 1-p & p \\ p & 1-p \end{bmatrix}$$

is given by $C = 1 - H(p, 1 - p)$

(ii) *Channles with Non-singular Channel Matrix :* For a square and non-singular channel P, the channel capacity is defined as:

$$C = \log_2 \sum_{j=1}^{n} \exp\left\{-\sum_{i=1}^{m} a_{jk} H(Y \backslash X = x_k)\right\}$$

where a_{jk} is the (j, k)th element of p^{-1}.

Efficiency

The efficiency of the noise-free system is given by, $\eta = \frac{H(X)}{L}$, where L is the length of the code.

Redundancy

The difference between the expected (or average) mutual information I(X, Y) and its maximum value is defined as the redundancy of the communication system. The ratio of given redundancy to channel capacity is known as relative redundancy, *i.e.,*

Redundancy for noise-free channel $\beta = C - I(X, Y) = \log n - H(X)$

Relative redundancy for noise-free channel β

$$= \frac{\log n - H(X)}{\log n} = 1$$

$$= \frac{H(X)}{\log n} = 1 - \text{Efficiency of the system.}$$

ENCODING

Encoding may be defined as a transformation procedure of a message from sources to receiver through a noiseless channel in some code language. In other words, if $X = \{x_1, x_2, \ldots, x_m\}$ be the set of messages to be transmitted, then codes may be defined as a relationship between all possible sequences of symbols of the set X with another set $Y = \{y_1, y_2, \ldots, y_n\}$ of code character of alphabet.

Objectives of Encoding

1. It is used to increase the efficiency of transmission.
2. It is used to minimize the expected code word length. If the code word associated with x_i is of length l_i, i = 1, 2, ... ,m, then the expected length of messages is given by

$$L = \sum_{i=1}^{m} l_i P\{x_i\}$$

The transmission for which L is minimum is considered to be efficient.

3. It is used to minimize the cost of transmission. If C_i, (i = 1, 2, ... ,m) is the cost of transmission of some words W_i with probability of transmission $P(W_i)$, then in a message of m words, the expected cost per message is given by

$$M = \sum_{i=1}^{m} C_i P(W_i)$$

The transmission for which M is minimum is considered to be efficient. Following are some of the subclasses of code:

(i) *Block Code* : A code which establishes a relationship with each of the symbols of the set X to a fixed sequence of symbols of the set Y is called a block code. That is, each symbol x_i (i = 1, 2, ..., m) is to be assigned a fixed sequence of symbols of Y called the *code words*, associated with x_i. For example, x_1 may correspond to y_1, y_2 and x_2 may correspond to y_7, y_8, y_4.

(ii) *Binary Code* : In particular, if the set X = {0, 1}, then a block code is said to be binary code. An example of binary block code is

$x_1 \rightarrow 0^1$, $x_2 \rightarrow 010$,

$x_3 \rightarrow 10$, $x_4 \rightarrow 110$.

(iii) *Non-Singular Code* : A block code is said to be non-singular code if all words of the code are distinct. An example of non-singular code is

$x_1 \rightarrow 00$, $x_2 \rightarrow 01$,

$x_3 \rightarrow 10$, $x_4 \rightarrow 11$.

(iv) *Uniquely Decodable (Separable) Code* : A code is said to be uniquely detonatable (separable) code if every finite sequence of symbols of the set Y is associated to at most one symbol of the set X. Examples of uniquely decodable (decodable) (separable) code are given below:

(a) $x_1 \rightarrow 0$, $x_2 \rightarrow 10$, $x_2 \rightarrow 110$, $x_3 \rightarrow 111$

(b) $x_1 \rightarrow 0$, $x_2 \rightarrow 01$, $x_3 \rightarrow 011$, $x_3 \rightarrow 0111$.

SHANNON-FANO ENCODING PROCEDURE

In this method a sequence of binary numbers (0, 1) is used for encoding messages through a memoryless communication channel. Let $X=\{x_1,x_2, ..., x_m\}$ be the list of the messages to be transmitted from some source and $P = \{p_1, p_2, ... ,p_m\}$ be their corresponding probabilities.

Our aim is to devise an encoding procedure so that a sequence of binary number (0, 1) of unspecified length can be associated to each message x_i. The sequence so obtained must satisfy the following conditions:

(i) No sequence of binary numbers can be obtained from any other sequence by adding additional binary terms to sequences of shorter lengths.

(ii) Binary numbers associated with each message x_i to form a sequence occur independently with equal probability.

The procedure can be summarized in the following steps:

Step 1:

Arrange the messages (words) $x_1, x_2, ...x_m$ in descending order in terms of their probabilities.

Without loss of generality, let $p_1 > p_2 ... > p_m$ so that, we have

Message : $x_1, x_2, ... ,xi, ... ,x_m$

Probability : $p_1, p_2, ... ,p_i, ... ,p_m$

Step 2:

Divide the set of messages $X = \{x_1, x_2, ... ,x_m\}$ into two subsets, say X_1 and X_2 of equal probabilities:

Set	*Message*	*Probabilities*
X_1	x_1, x_2	$P(X_1) = p_1 + p_2$
X_2	$x_3, x_4, ... ,x_m$	$P(X_2) = p_3 + ... + p_m$

i.e., $P(X_1) = P(X_2)$.

Step 3:

Again, divide both subsets X_1 and X_2 into two subsets, say X_{11}, X_{12} and X_{21}, X_{22} with equal probabilities respectively.

Step 4:

Assign binary number 0 to the first position of the coded word in each message in subset X_1 and binary number 1 to the first position of the coded word in each message in subset X_2.

The similar procedure of assigning binary number 0 and 1 must be repeated subsets of X_1 and X_2.

Step 5:

The division and assigning binary digits 0 and 1 continue till each subset contains only one message (word).

INFORMATION THEORY

The word *information* is very common in everyday language. Information transmission usually occurs through human voice (as in telephone, radio, television, etc.), books, newspapers, letters, etc. In all these cases a piece of information is transmitted from one place to another. However, one might like to quantitatively assess the quality of information contained in a piece of information.

(a) Suppose, one states that 'It is raining'. Now, the question is 'Have we received much information? Here it may be concluded that if a piece of information is presented which was already know, then, obviously, no information has been received. Again, if one states that 'The sun will shine the whole day informed that something will happen about which we did not know, therefore, we do not have to be surprised by the statement that was made.

(b) Suppose, we have come to know from the weather forecast on television that 'The rain will continue for the next 2 days.' In this case, we have received information and indeed more than in the second case in (a) because a statement has been made whole truth is not at all so surprising.

In the above two examples we have tried to relate the quantity of information with a prior likelihood of validity of the statements made. *Accor-ding to the usual way of looking at information, the quantity of information is aversely proportional tot hat likelihood.* This relationship may not truly describe the quantity of information in a statement about the likelihood of an event. But there are many measures which satisfy this relationship.

COMMUNICATION PROCESSES

The *communication process* may be defined as the procedure by which one mind affects the another. This may be any means by which the information is carried from a given source to the receiver. There are three essential parts of a communication system as explained below:

(i) *Source (or Transmitter)* : It is the source of message (either person or machine) which produces the information to be communicated or transmitted.

(ii) *Communication Channel* : It is the transmission network (or media) which carries the message from the source to receiver, *e.g.*, human voice, newspapers, books, etc. A communication channel can be with or without noise.

(iii) *Receiver* : It is the destination to which the message is conveyed from source (or transmitter) through a communication channel.

Other parts of the communication system are as follows:

(i) *Encoder* : It is an equipment which is used to improve the efficiency of the transmission channel through which a message is transmitted to the receiver.

(ii) *Noise* : It is the general term which creates interruptions or disturbances in the transmission of message (or information) from transmitter to receiver. For example, noise or disturbance in radio or television during the relay of a programme; error in newspaper printing, etc.

(iii) *Decoder* : It is used to transform encoded message into the original form at the receivers end.

The study of information theory is based upon a fundamental theorem which states that *It is possible to transmit information through a noisy channel at any rate less than the channel capacity with an arbitrarily small probability of error.*

The communication system described in Fig. 2.2 is statistical in nature because the source selects and transmits sequence of symbols from a given alphabet to the channel based on some statistical rule. The channel transmits this symbolic information to the receiver under some statistical rule also.

MEMORYLESS CHANNEL

A *memoryless channel* is described by an input alphabet $X = \{x_1, x_2, .., x_m\}$, an output alphabet $Y = \{y_1, y_2, ..., y_n\}$ and a set of conditional proba-bilities $p(y_j \mid x_i)$ of receiving the symbol y_j when symbol x_i is sent for all i and j.

If a memoryless channel is described only by two input symbols ($x_1 = 0$, $x_2 = 1$), two output symbols ($y_1 = 0$, $y_2 = 1$) and a set of conditional probabi-lities $p(y_j \mid x_i)$ for i, j = 1, 2, then it is called a *binary memoryless channel.*

A binary memoryless channel is always symmetric because

$$p(y_1 \mid x_1) = p(y_2 \mid x_1) = q$$
$$P(y_1 \mid x_2) = p(y_2 \mid x_2) = p$$

where $q = 1 - p$, p being the probability of error in transmission.

Remark : Since in this chapter we shall discuss only memoryless channel, therefore the word 'channel' will be used in place of 'Memoryless channel'.

The Channel Matrix

The input to the channel, the output form the channel and conditional probabilities for a pair of input symbol can be expressed in the form of a matrix called *channel matrix.*

$$\begin{array}{c} \text{Output, } Y \\ \text{Input, } X \begin{array}{c} \\ x_1 \\ x_2 \\ \vdots \\ x_m \end{array} \begin{array}{c} \begin{array}{cccc} y_1 & y_2 & \cdots & y_n \end{array} \\ \begin{bmatrix} p_{11} & p_{12} & \cdots & y_n \\ p_{21} & p_{22} & \cdots & p_{2n} \\ \vdots & \vdots & \vdots & \vdots \\ p_{m1} & p_{m2} & \cdots & p_{mn} \end{bmatrix} \end{array} \end{array}$$

where $p_{ij} = p(y_j \mid x_i)$; i = 1, 2, ... , m and j = 1, 2, ... , n.

The sum of conditional probabilities in each row must be equal to one.

Probability Relation in a Channel

If $p_{i0} = (x_i)$ denotes the probability that the symbol x_i is selected for transmission, $p_{0j} = p(y_j)$ the probability that the symbol y_j is received, then the relation between the probabilities of various input symbols and output symbols is expressed as:

$$\sum_{i=1}^{m} p_{i0} p_{j/i} = p0j;\ j = 1, 2, \ldots, n.$$

(i) The joint probabilities of sending a symbol x_i and receiving the symbol y_j is given by
$p(x_i, y_j) = p_{j/i}\ p_{i0}$ for all i.

(ii) The conditional backward input probabilities when it is known that the symbol y_j has been received is given by: $p(x_i \mid y_j) = p_{j/i}(p_{i0} \mid p_{0j})$ for all i and j.

Illustration : Consider a binary channel with input symbols X = {0, 1}, output symbols Y = {0, 1} and the channel matrix $\begin{bmatrix} 1/3 & 2/3 \\ 1/5 & 4/5 \end{bmatrix}$.

Further assume the input probabilities as: $p_{01} = 6/7$ and $p_{20} = 1/7$

Now, the output probabilities p_{01} and p_{02} can be obtained as follows by using the above stated relationship

$$p_{01} = p_{10}\ p_{1/1} + p_{20}\ p_{1/2} = \frac{6}{7}\times\frac{1}{3}+\frac{1}{7}\times\frac{1}{5} = \frac{11}{35}$$

$$p_{02} = p_{10}\ p_{2/1} + p_{20}\ p_{2/2} = \frac{6}{7}\times\frac{2}{3}+\frac{1}{7}\times\frac{4}{5} = \frac{24}{35}$$

The joint probabilities are obtained by using rule (i)

$$p(0, 0) = \frac{1}{3}\times\frac{6}{7} = \frac{2}{7} \qquad p(0, 1) = \frac{2}{3}\times\frac{6}{7} = \frac{4}{7}$$

$$p(1, 0) = \frac{1}{5}\times\frac{1}{7} = \frac{1}{35} \qquad p(1, 1) = \frac{4}{5}\times\frac{1}{7} = \frac{4}{35}$$

The conditional backward input probabilities are obtained by using rule (ii)

$$p(0 \mid 0) = \frac{1/3\times1/7}{11/35} = \frac{10}{11} \qquad p(0 \mid 1) = \frac{2/3\times6/7}{24/35} = \frac{5}{6}$$

$$p(1 \mid 0) = \frac{1/5\times1/7}{11/35} = \frac{1}{11} \qquad p(1 \mid 1) = \frac{4/5\times17}{24/35} = \frac{1}{6}$$

Noiseless Channel

If the channel matrix contains only one non zero element in each column, the such channel is called a noiseless channel. For example, the following channel matrix is a noiseless channel.

$$\begin{bmatrix} 1/2 & 0 & 1/6 & 0 \\ 0 & 5/7 & 0 & 0 \\ 0 & 0 & 0 & 2/3 \end{bmatrix}$$

A binary symmetric channel with probability, p = 0 or 1 is also a noiseless channel.

A MEASURE OF INFORMATION

Basic Assumptions : In order to illustrate a measure of information, the following assumptions are made:

(a) There is a finite set $X = \{x_1, x_2, \ldots, x_m\}$ of events and the probability of occurrence of event xi is pi (i = 1, 2, ... ,m), such that $p_1 + p_2 + \ldots + p_m = 1$. Consider that, the event x_k has occurred. So according to the statement that the quantity of information received is inversely proportional to the likelihood of the event, if $I(x_k)$ denote the amount of information received is inversely proportional to the likelihood of the event x_k, with probability p_k of occurrence, then $I(x_k) > I(x_r)$ for $p_k < p_r$.

(b) The amount of information received from the occurrence of x_k, $I(x_k)$ may be defined in several ways so as to satisfy the inverse relationship. For example, if $I(x_k) = \alpha/p_k$, $\alpha > 0$, then for each positive value of α, we may have several measures of $I(x_k)$. However, there are other properties also which are to be satisfied by a measure of information. For example, if an event, x_k has $p_k = 1$ of its occurrence, then $I(x_k) = 0$ because no information is received provided occurrence of a particular event is known in advance.

Whatever be the definition of $I(x_k)$, the expected value of information is given by

$$\sum_{i=1}^{m} p_i I(x_i)$$

If we are interested only in the probabilities of the occurrence of an event in set X and not in their actual natures, then the above expression for expected value of the information received may be written as:

$$\sum_{i=1}^{m} p_i I(p_i)$$

where $I(p_i) = -\log_2 p_i$. The $\log_2 p_i$ indicates the probability concerning the receiver before receiving the information provided the communication system is noiseless.

Choice of Measure : The expected value of information can also be interpreted as the *expected amount of information* needed to determine which event of set X has occurred. In other words, it is the measure of uncertainty about which event of X has occurred or will occur. The uncertainty is considered to be maximum when each event $x_1, x_2, \ldots x_m$ of X are equally probable, *i.e.*, when their probability of occurrence is equal,

$$p(x_1) = p(x_2) = \ldots = p(x_m) = 1/m.$$

Shannon and Wiener have suggested the following expression as the measure of expected amount of information

$$H(p_1, p_2, \ldots, p_n) = -\sum_{i=1}^{n} p_i \log_2 p_i$$

The function H is also known as the *entropy function.* Base 2 of the logarithm was a natural choice since information theory was initially concerned with communication and binary transmission. For example, let $X = \{x_1, x_2\}$ and $p(x_1) = p(x_2) = 1/2$. Then

$$H\left(\frac{1}{2},\frac{1}{2}\right) = -\left[\frac{1}{2}\log_2\left(\frac{1}{2}\right)+\frac{1}{2}\log_2\left(\frac{1}{2}\right)\right] = -\left[\frac{1}{2}(-1).2\right]$$

$$= 1 \text{ unit of information.}$$

This unit of information is the expected amount of information obtained due to occurrence of equally likely events x_1 and x_2. The name given to this unit of information is the bit.

PROPERTIES OF ENTROPY FUNCTION, H

1. *Continuity*

The entropy function, $H(p_1, p_2, \cdots, p_n)$ is continuous for each and every independent variable p_i; $0 \le p_i \le 1$.

$$H(p_1, p_2, \cdots, p_n) = p_i \log p_1 + p_2 \log p_2 + \ldots + p_n \log p_n$$

$$= p_1 \log p_1 + p_2 \log p_2 + \ldots + p_{m-1} \log p_m$$

$$+ (1 - p_1 - p_2 - \ldots - p_{n-1}) \log (1 - p_1 - p_2 - \cdots - p_{m-1})$$

Since all p_i, $i = 1, 2, \ldots, n$ are independent and continuous in the interval [0, 1], therefore logarithm of a continuous function will also be continuous.

2. *Symmetry*

The entropy function is a symmetric function in all variables. That is, H remains unchanged when p_i, i = 1, 2, ... ,n are interchanged with one another. Mathematically, it is stated as:

$$H(p_i, 1 - p_i) = H(1 - p_i, p_i);\ i = 1, 2, \dots n$$

3. *Maximum Value of H*

The entropy function H is a concave function because of the symmetric property.

$$H(p_i, 1 - p_i) = H(1 - p_i, p_i);\ i = 1, 2, \dots ,n$$

where $0 \leq pi \leq 1$. Now in order to maximize the information obtained from a single response, we should consider all possible subsets of set X of events with almost equal probability, *i.e.*, in H (p, 1 – p) if p cannot be equal to 1/2, then it should be very close to 1/2.

However, if response is of general nature, then divide the set X into k ($\leq$ n) subsets and obtain a response which identifies one of these subsets as that which contains the event in question. hence, the maximum value of H, *i.e.*,

$$\text{Max } H(p_1, p_2, \dots ,p_n) = \sum_{i=1}^{k} p_i \log p_i$$

$$\text{subject to} \quad \sum_{i=1}^{k} p_i = 1;\ p_i \geq 0$$

occurs for $p_1 = p_2 = \dots = p_k = 1/k$ and is given by – logk (1/k) = 1 unit. But a given unit is not 'bit' unless k = 2. Thus, the value of m should be as small as possible

4. *Additivity*

This property of H states that if a particular event x_n with probability p_n is divided into m mutually exclusive subsets say $e_1, e_2, \dots ,e_m$ with probabilities $q_1, q_2, \dots q_m$ respectively such that $p_n = q_1 + q_2 + \dots + q_m$, then

$$H(p_1, p_2, \dots ,p_{n-1}, q_1, q_2, \dots ,q_m)$$

$$= H(p_1\ p_2, \dots ,p_{n-1}, p_{n-1}, p_n) + p_n + H\left(\frac{q_1}{p_n}, \frac{q_2}{p_n}, \dots, \frac{q_m}{p_n}\right)$$

Adding and subtracting p_n log pn to the left hand side of additive property. Then we have

$$H(p_1, p_2, \dots, p_{n-1}, q_1, q_2, \dots, q_m) = \sum_{i=1}^{n-1} p_i \log p_i - \sum_{i=1}^{m} q_i \log q_i$$

$$= -\left\{\sum_{i=1}^{n} p_i \log p_i - p_n \log p_n\right\} - p_n \sum_{i=1}^{m} q_i \log q_i$$

$$= H(p_1, p_2, \dots, p_n) + \left\{p_n \log p_n - \sum_{i=1}^{m} q_i \log q_i\right\}$$

since $H(p_1, p_2, \dots, p_n) = -\sum_{i=1}^{n} p_i \log p_i$

But $p_n \log p_n - \sum_{i=1}^{m} q_i \log q_i = p_n \left\{\frac{p_n}{p_n} \log p_n\right\} - p_n \sum_{i=1}^{m} \left\{\frac{q_i}{p_n} \log q_i\right\}$

$$= \sum_{i=1}^{m} q_i \left\{\frac{p_n}{p_n} \log p_n\right\} - p_n \sum_{i=1}^{m} \left\{\frac{q_i}{p_n} \log q_i\right\}$$

$$= p_n \sum_{i=1}^{m} \frac{q_i}{p_n} \log p_n - p_n \sum_{i=1}^{m} \frac{q_i}{p_n} \log q_i$$

$$= -p_n \sum_{i=1}^{m} \frac{q_i}{p_n} (\log q_i - \log p_n)$$

$$= -p_n \sum_{i=1}^{m} \frac{q_i}{p_n} \log\left(\frac{q_i}{p_n}\right); \text{ since } p_n = \sum_{i=1}^{m} q_i$$

Now, $\text{LHS} = H(p_1, p_2, \dots, p_{n-1}, q_1, q_2, \dots, q_m)$

$$= H(p_1, p_2, \dots, p_n) + \left\{p_n \log p_n - \sum_{i=1}^{m} q_i \log q_i\right\}$$

$$= H(p_1, p_2, \dots p_n) - p_n \sum_{n=1}^{n} \frac{q_i}{p_n} \log\left(\frac{q_i}{p_n}\right)$$

$$= H(p_1, p_2, \dots p_n) + p_n H\left(\frac{q_1}{p_n}, \frac{q_2}{p_n}, \dots, \frac{q_m}{p_n}\right) = \text{RHS}.$$

NECESSARY AND SUFFICIENT CONDITION FOR NOISELESS ENCODING

Theorem 3:

(Noiseless Coding Theorem): *The necessary and sufficient condition for the existence of an irreducible noiseless encoding procedure with* *specified word length (n_1, n_2, ... ,n_N) and that a set of positive integers n_1, n_2, ... ,n_N can be found such that*

$$\sum_{i=1}^{N} D^{-n_i} \leq 1$$

where D is the number of symbols in encoding alphabet.

Proof:

Let x_i be the number of coded messages of length n_i. Since such messages having only letter cannot be greater than D, therefore $x_1 \leq D$.

Also due to the coding restriction, the number of message encoded of length 2 cannot exceed $(D - x_1)$ D, therefore we have

$$x_2 \leq (D - x_1) D = D_2 - x_1 D$$

Similarly, $$x_3 \leq \{(D - x_1) D - x_2\}D = D_3 - x_1 D_2 - x_2 D$$

$$\vdots$$

$$x_m \leq D_m - x_1 D^{m-1} - x_2 D^{m-2} - \ldots - x_{m-1} D$$

or $$x_m D_{-m} \leq 1 - x_1 D_{-1} - x_2 D_{-2} - \ldots - x_{m-1} D$$

or $$\sum_{i=1}^{m} x_i D^{-i} \leq 1 \qquad \ldots(7)$$

where m is the maximum length of any message.

The left hand side of inequality (7) can also be written as:

$$\sum_{i=1}^{m} x_i D^{-i} = x_1 D_{-1} + x_2 D_{-2} + \ldots + x_m D_{-m}$$

$$= \left[\frac{1}{D} + \frac{1}{D} + \ldots + x_1 \text{times}\right] + \left[\frac{1}{D^2} + \frac{1}{D^2} + \ldots + x_2 \text{times}\right] + \ldots + \left[\frac{1}{D^m} + \frac{1}{D^m} + \ldots + x_m \text{times}\right] \qquad \ldots(8)$$

Each term in the bracket of Eqn. (8) corresponds to a specified message length, such as in the first bracket, x_1 message is of length 1, in second bracket, x_2 message is of length 2 and so on. Hence the total number of messages are:

$$x_1 + x_2 + \ldots + x_m = N.$$

If $i = n_i$, then terms in Eqn. (11) can be rewritten as:

$$\sum_{i=1}^{m} x_i D^{-i} = \sum_{i=1}^{m} x_i D^{-i} \leq 1$$

This proves the necessary condition.

The proof of the sufficient condition is left as an exercise for the reader.

RESOURCE SCHEDULING

During the developing of PERT and CPM networks, we have generally assumed that sufficient resources are available to perform the various activities. At a certain time the demand on a particular resource is the cumulative demand of that resource on all the activities being performed at that time. Going according to the developed plan, the demand on a certain type of resource may fluctuate from very high at one time to a very low at another.

If it is a material or unskilled labour which has to be procured from time to time, the fluctuation in demand and will not much, affect the cost of the project. But if it is some personnel who cannot be hired and fired during the project or machines which are to be hired for the total project duration, the fluctuation in their demand will affect the total project cost due to high idle time. To reduce the idle period, the activities on non-critical paths are shifted by making use of the floats, and an alternate schedule is generated comparing the important resources, with the object of smoothening the demand on resources.

In some situations we may be faced with a demand for some critical resource which may be limited in supply. For example, the only bulldozer available may be needed for two activities at two places at a time. This makes the schedule infeasible and calls for a re-examination with the object of generating an alternate plan with feasible scheduling of the limited resource.

Thus, the object of resource scheduling is two-fold : it aims at bringing down the costs and at the same time reduces pressure on the limited resources conflicting-demands.

Depending upon the type of constraint, the resource scheduling situation may of two types:

(a) The constraint may be the total project duration. In this case the resource scheduling only smoothens the demand on resources in order that the demand of any resource is as uniform as possible.

In this case the resource scheduling is called 'Resource Smoothening' or 'Load Smoothening'.

(b) The second type of constraint may be on the availability of certain resources. Here the project duration is not treated as an invariant, but the demand on certain specified resources should not go beyond the specified level. This operation of resource allocation is called '*Resource Levelling*' or '*Load Levelling*'.

RESOURCES LEVELLING

The analysis aiming at stabilization of rate of resources utilization by various activities at different times without changing the project duration is called resources levelling.

In order to stabilize the use of existing level of resources that total float of non-critical activities is used. By shifting a non-critical activity between its earliest start time and latest allowable time, project manager may be able to lower the maximum resources requirement.

The following two general rules are normally used in scheduling non-critical activities:

(i) If the total float of non-critical activity is equal to its free float, then it can be scheduled anywhere between its earliest start and latest completion times.

(ii) If the total float of non-critical activity is more than its free float, then its starting time can be delayed relative to its free float, then its starting time can be delayed relative to its earliest start time by no more than the amount of its float without affecting the scheduling of its immediately succeeding activities.

RESOURCES SMOOTHING

The analysis aiming to reduce peak demand for resources and re-allocating among activities of a project in a manner so that the total projects duration remains shortest is known as resources smoothing (or loading).

The procedure of carrying out resource smoothing can be summarized in the following steps:

Step I. Calculate the earliest start and latest finish times of each activity and then draw a time scaled version of the network. In this network critical path is drawn along straight line and non-critical activities on both sides of the line. Resource requirement of each activity is given along the arrows.

Step II. Draw the resource histogram by taking earliest start times or latest start times of activities on the x-axis and cumulative resources required on y-axis.

Step III. Shift start time of non-critical activities from having largest float in order to smoothen the resources.

COMPLEXITIES OF PROJECT SCHEDULING WITH LIMITED RESOURCES

The process of load smoothening and load levelling have been demonstrated with the help of a small project comprising of only 11 activities, and a resource of only one type. The activities were shifted from high-load periods to low-load periods just arbitrarily to smooth or level the resource requirement. In this small network, the juggling of activities was possible.

From this example, we observe that in levelling of load to which we resort in case of limited resources, the job start are constrained not only by their precedence relationship, but also by the availability of the resource in the levelled network, generally, there is no critical path from the activity duration point of view. Thus, the concepts of slack and criticality lose their normal meanings.

In a real life project, a number of resources are required in the form of men, machines and materials, and many of these are available only in limited quantities. The problem of scheduling a project with multiple of limited resources is a very different task, even for a project of moderate size. A very large number of feasible schedules are possible in such a project. To determine all these schedules and then to find the best out of these, by the method of juggling activities, is just impossible such a problem, where a large number of feasible combinations of job start times are possible, is called a *combinational problem*. No mathematical technique is yet available to solve such a problem of resource scheduling. Integer Programming (I.P.) has been employed by some O.R. analysts, but with a limited success, only to small projects. For large projects the I.P. also becomes computationally impracticable.

Only the *heuristic methods* have been found helpful in such large combinatorial problems. The heuristics, which in simple terms is just a set of rules, may not lead to an optimal solution, but definitely lead to a practical solution close to optimal and with a moderate effort on the part of the analyst.

A heuristic a collection of rules for solving a problem, is not an analytical technique, but a sort of guide for accomplishing the objectives. In a small problem, manual application of the heuristic technique may be possible, but the large projects require the employment of computers. A heuristic procedure put in the form of a computer program is called *heuristic program.* A number of such programs for project scheduling with limited resources have been developed by different persons. Here we will discuss the heuristic program for resource levelling, when there is no constraint on project duration. The object is to find the shortest project schedule subject to the constraint on resource availability. This may be called a *resource allocation problem.*

PROGRAMMES FOR RESOURCE ALLOCATION

The best method of illustrating the heuristic programmes is with the help of some examples.

Consider the project the required data for which is given in Table 4.2. After determining the critical path, the project has been drawn as a squared network using early start schedule . Only one type of resource, the manpower required for executing different activities, has been considered in this case. The daily manpower required for each activity has been shown along the activity arrow. The daily manpower required for executing the project accor-ding to the early start schedule has also been shown below the squared network. Now, there let be 10 men available for the project. The project will have to be rescheduled, so that the demand on men is always less than or equal to 10, and at same time the project duration is extended to the minimum possible. This is the case of load levelling. A number of heuristic procedures are available for tackling such problems a common on being the day-by day scheduling. This program is based on the following rules (heuristic):

1. Starting with day one, allocate resource serially in time.
2. When a number of activities complete for the same resource assign priority according to the criticality of the job, job with least slack is taken up first.
3. Reschedule non-critical jobs, if possible, when resources are required fir critical or non-slack jobs.
4. Jobs once started are not interrupted.

PERT/COST, DECISION CPM, PRECEDENCE NETWORKING AND GENERAL REMARKS

PERT/Cost

The PERT/Cost system was developed in 1962, by the U.S. Department of Defence and the National Aeronautics and Space Administration; and was published in a U.S. Government manual, 'DOD and NASA Guide; PERT Cost System Design.' The basic concept of PERT and CPM cost system or simply PERT/Cost is that costs are measured and controlled on a project basis, rather than according to functional organisation of the firm. Instead of departments of the organisation, it is the activities or work packages (groups of activities), which form the basis of project control, in the PERT/Cost system. PERT/Cost integrates the time data, the cost data and the work accomplished. The essence of PERT/Cost system is that the responsibility for expenditure should coincide with the responsibility of managing the associated work. In this process, PERT/Cost aims at achieving a more realistic estimate of the project cost and to have a better control of costs when the project is in progress. Cost overruns are easily Y defected and corrective actions readily taken by controlling the activities responsible for over-runs.

Work Packages

As discussed above, the basic idea of PERT/Cost system is that the activities, which form the basis for project planning and scheduling. Should also form the basis for the measurement and control of costs. The activities are the result of work breakdown structure, which could be either end-oriented or functionally oriented, that is organisationally structured. The size of an activity depends upon the level of work breakdown, which further depends upon a number of factors. In many cases, the activities into which a project is broken down, may be essential for the purpose of detailed planning and scheduling, but may be too small for the purpose of cost accounting and control. In such a case, the activities generally falling on a continuous path and related to each other are grouped together to form a cost work package or simply work package. The work package which may comprise of a single activity or a group of activities, represents a particular unit of work, for which the responsibility of execution and cost control can clearly be defined, and which is manageable for planning and control purposes. It has a clear starting and terminating event and is preferably single functional. The formation of work packages from a detailed network. While forming a work, package care should be taken that the activities being grouped arc related to each other from the accounts point of view as well as from the execution point of view, that is the precedence of activities should be taken in to consideration. A work package in PERT/Cost system is identified by a charge account number in the cost chart and is generally the responsibility of one individual.

The level of accuracy in PERT/cost estimates depends to a large extent no the size of work packages, which in turn depends upon a large number of parameters, such as size and complexity of the project, type of industry, the detail obtained in W.B.S. and the stage of the project. When the project is of huge size, the work packages are also generally large. When a project is blown up into greater details, the work packages arc comparatively small in size. At certain stages of project, more detailed information i.e., smaller work packages may be required. The type of industry to which the project belongs also affects the size of the work packages. Work packages may be large in the aero-space craft and ship building industry, as compared to those in an electronic industry.

Though, the factors discussed above affect the size of the work packages, work are no specific rules or guide lines which can be used in the formation of work packages in a project. The analyst is required to look into the various aspects of the project, the structure of the organisation, the level of management and make best use of his judgement, while grouping the activities into work packages.

The DOD and NASA Guide, written primarily for the aero-space industry, suggests the value of a work package to be no more than $ 1,00,000 and of duration no more than 3 months. On the other hand in a project on the development of Boeing 707, none of the work package required more than 1,000 man-hours and none required more than 15 weeks execution time.

PROJECT COST SCHEDULE

The cost schedule of a project is prepared by adding the cost of work packages, period by period, according to the activity time schedule. The activity time schedule may be based on the early start or laic start of the activities, or on any other feasible activity schedule. While computing the project schedule from period to period, it is assumed that the expenditure on a work package is made at a constant rate over the duration of the work package. In some cases, this assumption may be very severe. Such work packages should be divided into sub-work packages, each of which has a relatively constant expenditure rate.

The early start times, the late start times, activity or work package durations and total cost per week of the work-packages are given in Table 4.3. The scheduled completion time for the project has been taken to be equal to the earliest completion time which is 22 weeks. The same network as a schedule graph. Two schedule networks are drawn, one for the early start schedule and the other for the late start schedule. The cost per week has been computed for each week, by adding the cost of works 10 be carried in that week. For example, in the first weak of early start schedule, activities 1-2 and 1-3 are to be carried out at a cost of Rs. (1.000 + 1,500) = Rs. 2,500, while in the 8th week of the same schedule, four activities 5-7, 5-6, 2-4 and 3-4 are active, giving the cost for that week as Rs. 3,800 [Rs. (13.00 + 1.000 + 1,200 + 400)]. Along with the cost per month, the cumulative-costs have also been given find two schedules.

Table 1

Activity or Work Package	*Duration (Weeks)*	*Early Start Time*	*Late Start Time*	*Total Cost (Rs.)*	*Cost Per Week (Rs.)*
1-2	4	0	0	4,000	1.000
1-3	3	0	8	4,500	1,500
2-4	5	4	11	6,000	1,200
2-5	3	4	4	4,500	1,500
3-4	5	3	11	2,000	400
4-6	2	9	16	1,00	800
5-6	7	7	11	7,000	1,000
5-7	5	7	7	6,000	1,200
6-8	2	14	18	1,000	500
7-8	8	12	12	12,800	1,600
7-9	6	12	16	4,200	700
8-9	2	20	20	2,000	1,000

In cumulative costs for the early son and late start schedules have been illustrated graphically. The area between the two curves represents a range of feasible project cost schedules. This graphical representation gives a clear idea of the budget implications of the early start and late start schedules. From the budgetary point of view, the cumulative project cost should follow a relatively straight line *i.e.*, the expenditure should not vary much from period to period. This can be approximated by shifting the activities on the non-critical path to smoothen the costs. Fig. 4.9 shows one such, schedule for which the cost curve in Fig. 4.8 is nearly a straight-line.

PROJECT MONITORING AND COST CONTROL

An essential requirement of project management is the mechanism for continuous comparisons of actual expenditure made to the emanated expenditure.

The PERT/Cost System Acts as Such a Mechanism

In addition to comparison of costs, it provides a means for comparing work scheduled and the work accomplished. As the project progresses, at various points in time, which we may call milestones, the management would like to check the slate of the project, that is whether it is on schedule, what is the new expected completion time, the cost incurred on the project *vis-a* the estimated cost, the work accomplished, delays and overruns—their causes and sources and the like. This detailed information helps the management to initiate, appropriate action to control the performance and cost of the project The state of the project can more easily be maintained by graphic aids. Curve I depicts the budgeted cost of the project obtained during the planning phase. This curve corresponds to the accepted feasible schedule between the early start and late start schedules. The actual cost incurred in the execution of the project is shown by curve II in the figure.

A simple comparison of the expected and actual cost curves can not reveal overrun or short runs, i.e. whether the actual expenditure is over or short of the estimated. It depends upon the extent of work; accomplished by that date. The value of work done is given by curve III. Let us check the state of the project at the end of 8th week. The estimated project cost is Rs. 20,600, while the actual cost as given by curve is Rs. 25,000. This .gives the impression that the expenditure made on a project upto this dale is more than required (estimated).

Activity 2–4 has already been completed, activity 3–4 needs 2 weeks more, 50% work on activity 5–7 has been completed and activities 4–

6 and have yet to start. Knowing the state of the project, the value of work accomplished is obtained as Rs. (17,800 + 3,000 + 2,400) = Rs. 23.200, which is less than the expenditure made.

$$\text{The cost overrun (underrun)} = \frac{\text{Actual cost - value of work done}}{\text{Value of work done}}$$

The positive value indicates overrun, while the negative value gives underrun. In the present case there is overrun of about 7.8% at the end of 8th week. The project, however, is ahead of schedule. The work worth Rs. 23,200 should have been completed near the end of 9th week. Thus the project is ahead of schedule by one week, while the expenditure is Rs. 1,800 more than the value of work completed. Curve IV shows the % overrun, while curve V shows the duration by which the project is ahead or behind the schedule. This information about the project, status helps the management to apply corrective measures to control the project. The forecast of the future performance can be made by projecting these trends into future. Depending upon the state of the project, management may decide to revise the cost estimates and/or the project schedule.

To exercise effect live control of the project, it is very essential to analyse the status and to find the causes and sources of cost and schedule problems. The cost and performance curves which are based on the total project, do not provide the necessary insight for this purpose. Similar curves based sub-divisions of the project or departments of the organisation, if drawn, would provide the useful information. These cost curves may not pin point the causes of overruns or delays, but will definitely point towards the sources and causes of project cost and schedule troubles. These curves will help to identify the sub-divisions or departments of the project, which have problems and need better control.

DECISION CPM

The network techniques of PERT and CPM specifically differentiate the managerial functions of planning and scheduling. The activities to be per-formed, their durations, resource requirements and methods to be adopted for accomplishing the etc. are determined at the planning stage. Scheduling, on the other hand concerns with the laying out of actual activities of the project in logical sequence, allocation of resources to jobs and expected completion times of the jobs, etc. In each project there are some activities which can be performed by alternative methods or there may be alternatives to the activities themselves. The selection of an alternative at one stage may affect the next activities to be taken up or may affect the total project duration and cost In the conventional CPM effort is made to select the best out of the available alternatives, but the effect of this choice on the future activities and the schedule of the project is not analysed. This practice, however, has been criticised by some observers. Since the selection of an alternative job or method is likely to affect the scheduling of the project, it is argued that the scheduling implications of a decision be considered before making it final. Decision CPM.

Which is the outcome of these arguments, continues the of concept of a decision tree with the AON diagram. The Decision network comprises of two types of nodes, activity nodes and decision nodes. This extremely small project having 6 alternative project networks can be analysed by hand computations by the method of enumeration.

But in a project of even moderate size, say, having 10 decision nodes with three alternatives at each node, the number of alternative networks would be $3^{10} = 59049$. Thus it will become, if not impossible, extremely uneconomical to enumerate all the possible combinations of alternatives. In a real life project comprising of activities in hundreds, with many as decision nodes having multiple alternatives, even high speed computers will refuse to enumerate all the possible networks.

Such combinatorial problems arc generally solved by some heuristic techniques or by employing the branch and bound technique. Such techniques help to reduce the number of alternatives to be evaluated. Crowston and Thompson, in their paper on Decision CPM, have presented one heuristic procedure to analyse such combinatorial project networks. As is true with all heuristics, this approach does not ensure an optimal solution to the problem, but promises a solution reasonably good as compared to the computational effort employed.

How the Networks (PERT/CPM) Help Management

1. The networks clearly designate the responsibilities of different supervisors. Supervisor of an activity knows his time schedule precisely and also the supervisors of other activities with whom be has to co-ordinate.
2. These techniques help the management in achieving the objective with minimum of time and least cost and also in predicting the probable duration and the associated cost.

3. The networks provide a number of checks and safeguards against going astray in developing the plan for the project and thus there are little chances of oversight of certain activities and events.
4. The flexibility of the networks permits the management to make the necessary alterations and refinements as and when needed. These allocations may be made during deployment of resources or reviewing.
5. The network techniques help the management in planning the complicated projects, controlling the implementation of the plan and keeping the plan up-to-date. They also help in locating the potential trouble spots and in taking corrective measures.
6. Application of network techniques has resulted in better managerial control, improved utilisation of resources, improved communication and progress reporting and better decision making.
7. Applications of PERT and CPM have resulted in saving of time which directly results in saving of cost. Saving in time or early completion of the project results in earlier return of revenue and introduction of the project or process ahead of the competitors resulting in increased profits.

Few Comments on the Assumptions of PERT and CPM

1. It is not always possible to sort out completely identifiable activities and their start and finish times.
2. Cost-time trade-offs used for finding the cost curve slopes are also subjective and a great effort and expertise is required to estimate them.
3. Beta distribution may not always be applicable.
4. Time estimates have an element of subjectiveness in them.
5. In calculating the S.D. of the critical path, independence of activities is assumed. Limitations of resources may invalidate this independence of activities.
6. The formulae for the expected duration and standard deviation are simplifications. In certain cases the errors, due to these assumptions, may even be of the order of 33%.

Difficulties in Using Network Methods

Following are some of the problems faced in the managerial use of network methods:

1. The planning and implementation of networks require personnel trained in the network methodology. Managements are reluctant to spare the existing staff to learn these techniques or to recruit trained personnel.
2. Determination of the level of network detail is another troublesome area. The level of detail varies from planner to planner and depends upon the judgement and experience.
3. Difficulty in securing the realistic urne estimates. In the case of new and non-repetitive type projects, the time estimates produced are often mere guesses.
4. Developing a clear logical network is also troublesome. This depends upon the data input and thus, the plan can be no better than the personnel who provides the data.
5. The natural tendency to oppose changes results in the difficulty of persuading the management to accept these techniques.

Applications of Network Techniques

The list containing PERT and CPM applications is very large and the applications arc expanding to many new areas. Following are a few typical areas in which these techniques are widely accepted :

1. *Administration :* Networks have been used by the administration for straitening paperwork system, for making major administrative system revisions, for long range planning and developing staffing plans, etc.
2. *Marketing :* Networks have been used for advertising programmes for development and launching of new projects and for planning their distribution.
3. *Construction Industry :* It is one of the largest areas in which the network techniques of project management have found application.
4. *Research and Development :* R and D has been the most extensive area where PERT has been used for development of new products, processes and systems.
5. *Inventory Planning :* Installation of production and inventory control, acquisition of spare parts, etc. have been greatly helped by network techniques.

6. *Maintenance Planning :* Shutdown and Maintenance of power plants, chemical plants, steel furnaces and overhauling of large machines have been carried out by using PERT.
7. *Manufacturing :* The design, development, and testing of new machines, installing machines and plant layouts are a few examples of how it can be applied to the manufacturing function of a firm.

SOLVED EXAMPLES

Example 1:

Evaluate the average uncertainty associated with the sample space of events A, B and C which are mutually exclusive with probability distribution

Event	:	*A*	*B*	*C*
Probability	:	*1/5*	*4/15*	*8/15*

Solution:

From the data of the problem, we have

$$p_1 = 1/5,\ p_2 = 4/15 \text{ and } p_3 = 8/15$$

The entropy function H is defined as

$$H(p_1, p_2, \ldots, p_n) = -\sum_{i=1}^{n} p_i \log p_i$$

where p_i's are the probabilities associated with the given probability distribution. For the given example n = 3, therefore

$$H(1/5, 4/15, 8/15) = -p_1 \log p_1 - p_2 \log p_2 - p_3 \log p_3$$

$$= -\frac{1}{5}\log\left(\frac{1}{5}\right) - \frac{4}{15}\log\left(\frac{4}{15}\right) - \frac{8}{15}\log\left(\frac{8}{15}\right)$$

$$= -\frac{1}{15}\left[3\log\frac{1}{5} + 4\log\left(\frac{4}{15}\right) + 8\log\left(\frac{8}{15}\right)\right]$$

$$= -\frac{1}{15}[-3\log 5 + 4(\log 4 - \log 15) + 8(\log 8 - \log 15)]$$

$$= -\frac{1}{15}[-3\log 5 - 4\log(3\times 5) - 8\log(3\times 5) + 4\log(2)^2 + 8\log(2)^3]$$

$$= \frac{1}{15}[15\log 5 + 12\log 3 - 32];\ \log 2 = 1$$

$$= \log 5 + \frac{4}{5}\log 3 - \frac{32}{15}$$

Example 2:

Show that the entropy of the following probability distribution is

$$2 - \left(\frac{1}{2}\right)n - 2$$

Event	:	x_1	x_2	...	x_i	...	x_{n-1}	x_n
Probabilities	:	$\frac{1}{2}$	$\frac{1}{2^2}$	...	$\frac{1}{2^i}$	...	$\frac{1}{2^{n-1}}$	$\frac{1}{2^n}$

Solution:

From the data of the problem, we have

$$p_i = \frac{1}{2^i},\ i = 1, 2, \dots, n - 1;$$

and $p_n = \frac{1}{2^{n-1}};\ \sum_{i=1}^{n} p_i = 1$

The entropy function H is defined as

$$H(p_1, p_2, \dots, p_n) = -\sum_{i=1}^{n} p_i \log p_i = -\sum_{i=1}^{n-1} p_i \log p_i - p_n \log p_n$$

$$= -\sum_{i=1}^{n-1}\left(\frac{1}{2^i}\right)\log\left(\frac{1}{2^i}\right) - \left(\frac{1}{2^{n-1}}\right)\log\left(\frac{1}{2^{n-1}}\right)$$

$$= \sum_{i=1}^{n-1}\left(\frac{1}{2^i}\right)\log(2^i) + \left(\frac{1}{2^{n-1}}\right)\log_2(2^{n-1})$$

$$= \sum_{i=1}^{n-1} i\cdot\left(\frac{1}{2^i}\right) + (n-1)\left(\frac{1}{2^{n-1}}\right)\quad [\log_2(2) = 1]$$

$$= \left\{\frac{1}{2} + \frac{2}{2^2} + \frac{3}{2^3} + \dots + \frac{n-1}{2^{n-1}}\right\} + \frac{n-1}{2^{n-1}} \qquad \dots(9)$$

or

$$\frac{1}{2}H(p_1, p_2, \dots, p_n) = \left\{\frac{1}{2^2} + \frac{2}{2^3} + \frac{3}{2^4} + \dots + \frac{n-1}{2^n}\right\} + \frac{2-1}{2^n} \qquad \dots(10)$$

Subtracting eq. (10) from eq. (9), we get,

$$H(p_1, p_2, \ldots, p_n) - \frac{1}{2}H(p_1, p_2, \ldots, p_n)$$

$$= \left\{\frac{1}{2}+\frac{1}{2^2}+\frac{1}{2^3}+\ldots+\frac{1}{2^{n-1}}\right\}+\left\{\frac{n-1}{2^{n-1}}-\frac{2(n-1)}{2^n}\right\}$$

or $$\frac{1}{2}H(p_1, p_2, \ldots p_n) = \left\{\frac{1}{2}+\frac{1}{2^2}+\frac{1}{2^3}+\ldots+\frac{1}{2^{n-1}}\right\} = 1 - \left(\frac{1}{2}\right)^{n-1}$$

or $$H(p_1, p_2, \ldots, p_n) = 2 - \left(\frac{1}{2}\right)^{n-2}$$

Example 3:

A transmitter has a character consisting of five letters (x_1, x_2, x_3, x_4, x_5) and the receiver has a character consisting of four letters, (y_1, y_2, y_3, y_4). The joint probability for the communication is given below:

$P(x_i, y_j)$	y_1	y_2	y_3	y_4	$P(x_i)$
x_1	0.25	0	0	0	0.25
x_2	0.10	0.30	0	0	0.40
x_3	0	0.05	0.10	0	0.15
x_4	0	0	0.05	0.10	0.15
x_5	0	0	0.05	0	0.05
$P(y_j)$	0.35	0.35	0.20	0.10	—

(a) *Determine the different entropies for the channel, assume that $0 \log 0 = 0$.*

(b) *Determine $H(X)$, $H(Y)$, $H(X, Y)$ and $H(Y \mid X)$.*

Solution:

In order to determine different entropies for the channel, joint probabilities have to be calculated for values of and j. Then different marginal and conditional probabilities may be calculated with the help of joint probabilities as given below:

$$P(x_1) = 0.25 + 0.0 = 2.25$$

$$P(x_2) = 0.10 + 0.30 = 0.40$$

$$P(x_3) = 0.05 + 0.10 = 0.15$$

$$P(x_4) = 0.05 + 0.10 = 0.15$$

$$P(x_5) = 0.05 + 0.0 = 0.05.$$

Similarly, $P(y_1) = 0.35$, $P(y_2) = 0.35$, $P(y_3) = 0.20$, $P(y_4) = 0.10$

The conditional probabilities $P\{x_i \mid y_j\}$ may then be calculated as

$P(x_i \mid y_j)$	y_1	y_2	y_3	y_4
x_1	1	0	0	0
x_2	0.25	0.75	0	0
x_3	0	0.33	0.66	0
x_4	0	0	0.33	0.66
x_5	0	0	1	0

Marginal Entropies

$$H(X) = -\sum_{i=1}^{5} P\{x_i\} \log P\{x_i\}$$

$$= (0.25) \log (0.25) - (0.40) \log (0.40) - (0.15) \log (0.15) - (0.15) \log (0.15) - (0.05) \log (0.05)$$

$$= -\frac{1}{4}\log\left(\frac{1}{4}\right) - \frac{2}{5}\log\left(\frac{2}{5}\right) - \frac{3}{10}\log\left(\frac{3}{20}\right) - \frac{1}{20}\log\left(\frac{1}{20}\right)$$

$$= \frac{1}{4}\log 4 + \frac{2}{5}\log\left(\frac{5}{2}\right) + \frac{3}{10}\log\left(\frac{20}{3}\right) + \frac{1}{20}\log 20 = 1.326 \text{ bits}$$

$$H(Y) = -\sum_{j=1}^{4} p\{x_j\} \log P\{y_j\}$$

$$= -(.35) \log (0.35) - (0.35) \log (0.35) - (0.20) \log (0.20) - (0.10) \log (0.10)$$

$$= \frac{7}{10}\log\left(\frac{20}{7}\right) + \frac{1}{5}\log(5) + \frac{1}{10}\log 10 = 1.855 \text{ bits}$$

Conditional Entropies

$$H(X \mid Y) = -\sum_{i=1}^{5}\sum_{j=1}^{4} P\{x_i, y_j\} \log P\{x_i | y_j\}$$

$$= -\left\{(0.25)\log\left(\frac{0.25}{0.35}\right) + 0.10\log\left(\frac{0.10}{0.30}\right) + 0.30\log\left(\frac{0.30}{0.35}\right)\right.$$

$$+ 0.05\log\left(\frac{0.05}{0.35}\right) + 0.10\log\left(\frac{0.10}{0.20}\right) + 0.05\log\left(\frac{0.05}{0.20}\right)$$

$$+0.05\log\left(\frac{0.05}{0.20}\right)+0.10\log\left(\frac{0.10}{0.10}\right)\Bigg\}$$

$$= -\left\{\frac{1}{4}\log\left(\frac{5}{6}\right)+\frac{1}{10}\log\left(\frac{1}{3}\right)+\frac{3}{10}\log\left(\frac{6}{7}\right)+\frac{1}{20}\log\left(\frac{1}{7}\right)\right.$$

$$\left.+\frac{1}{10}\log\left(\frac{1}{2}\right)+\frac{1}{20}\log\left(\frac{1}{4}\right)+\frac{1}{20}\log\left(\frac{1}{4}\right)+\frac{1}{10}\log(1)\right\} = 0.0704$$

bits

$$H\,(Y \mid X) = H\,(Y) + H\,(X \mid Y) - H\,(X)$$

$$= 1.855 + 0.0704 - 1.326 = 0.599 \text{ bits}$$

Joint Entropy

$$H\,(X, Y) = H\,(X) + H\,(Y \mid X) = 1.326 + 0.599 = 1.925$$

Example 4:

The resource requirement (man-days) and the range of permissible crew size for the activities of a project are given in Table 1. Any crew size with in the permissible range may be employed to carry out the job, subject to the condition that the total number of men employed on any day should not exceed 16. Schedule the project to minimise the project duration.

Table 1

Job	***Immediate Predecessor (Man-days)***	***Resource Required***	***Range of Permissible Crew Size***	
			Minimum	***Maximum***
a	–	*45*	*5*	*15*
b	*a*	*24*	*2*	*6*
c	*a*	*96*	*4*	*12*
d	*a*	*50*	*2*	*10*
e	*d*	*30*	*2*	*10*
f	*c*	*60*	*4*	*12*
g	*c*	*12*	*1*	*6*
h	*b*	*35*	*5*	*7*
i	*f*	*42*	*3*	*14*
j	*g, e*	*24*	*3*	*8*
k	*i, j*	*20*	*2*	*10*

Solution:

It may be started with a network based on maximum crew size and then expanding the activities while allocating the resources or with a network based on minimum crew size and crashing the jobs during allocation of resources.

Since the normal times are not given the following procedure has been used to find the pseduo-normal durations of jobs.

Total man-days required in the project are 438. Since 16 men are to be employed, length of the critical path should be a minimum of 438/16 $\simeq$ 28 days. The network, based on largest crew size (minimum activity duration) gives a critical path of 21 days. Thus each activity on the critical path be increased by multiplying the minimum duration by 23/21 and then rounding it off to the nearest acceptable whole number. For example, the duration of job A can be 45/5 = 9, 45/ 9 = 5, 45/15 = 3 days and can not be 3 $\times$ 28/21 = 4 days. Acceptable duration will the 5 days (crew size of 5). Similarly, for activity C, the normal duration 96 $\times$ 28/21 = 10.6 can be rounded to the nearest acceptable duration of 12 days (crew sizc of 8).

The network is then drawn using those normalised durations for the critical activities. Activities on the non-critical paths are then expanded to fill the slack. The slack on a path is distributed between the contending activities in the ratio of their durations. The schedule graph with activities expanded. Now the day-by-day resource scheduling resorted to.

Day 1 : There is only one activity A, and the same can be scheduled at the maximum crew size of 15 men.

Days 2 and 3 : Activity A with 15 men is continued.

Day 4 : Activities B, C and D can start. The load as computed is 15 men. The remaining one man should also be allocated if possible. Starting with critical activities, addition of one man to C is not possible, neither is it possible to add one man to D. It can be added to allocated manpower of B to make it 3 (activity duration 8 days).

Days 5 to 11 : Continue with the schedule of day 4.

Day 12 : Priority is given to jobs already in progress.

Continue activity C with 8 men.

Continue activity D with 5 men.

Activity H can be started with the remaining 3 men. Since H has slack, it can be postponed.

Day 13 : Continue C and d with 13 men postpone H.

Day 14 : Continue activity C with 8 men.

Schedule critical activity E with 3 men.

Since H can not be scheduled with remaining 3 men, postpone it.

Now check, if the remaining three men can be allocated to jobs C and E. Since, the next crew size of C is 12, none can be added to the present crew size of 8. Crew size of activity E can be increased to 6 to reduce its duration to 5 days.

Day 15 : Continue jobs C and E with 14 men.

Day 16 : Continuing the job E, already in progress with 6 men, activities F, G and H are due to start on this day. Activities F and G are critical and require 13 (10 + 3) men. One of these will have to be extended. Next possible crew size for F is 6. After allocating 6 men to E, 6 to F and 3 to G, one man remains. With one man, H can not stand, postpone it and add one man to the crew of activity G.

Days 17 to 18 : Continue the jobs E, F and G with 16 men, and postpone H.

Day 19 : Continue job F with 6 men.

Schedule job J with 3 men.

Schedule job H with 5 men.

Out of the remaining 2 men, none can be allocated 10 F and only one can be added to the crew of J. The remaining one can not be added to the crew of H and will thus remain idle.

Days 20 to 24 : Continue with the schedule of day 19th. Continue F with 6 men and H with 5 men.

No other activity is due for scheduling on this day.

Day 26 : There is only one activity I, requiring a maximum of 16 men. Schedule it

Day 27 & 28 : Job I is continued.

Day 29 : There is only one activity K. Schedule it employing the maximum permissible crew size of 10.

Day 30 : Job K is continued.

The project duration is 30 days. Now the amount of idle man-days can be computed by summing up the idle man-days for the entire project duration of 30 days which comes to 42 man-days.

Example 5:

The activities, their durations and skilled and unskilled men required their execution are given below.

Activity		*A*	*B*	*C*	*D*	*E*	*F*	*G*	*H*	*I*	*J*
Duration (days)	:	*1*	*4*	*3*	*1*	*2*	*2*	*2*	*3*	*1*	*3*
Skilled men reqd. day	:	*7*	*1*	*4*	*3*	*5*	*3*	*2*	*6*	*7*	*10*
Unskilled men reqd. day	:	*12*	*3*	*9*	*6*	*10*	*8*	*4*	*10*	*14*	*15*

Solution:

The precedence relationships are : B, C, D, E, F > B; G > C; H > D; I > G, H; J > I, E.

The project is to be scheduled for the case when only 10 skilled and 20 unskilled men arc available throughout the project duration.

The values along the activity arrows represent the daily resource requirement as skilled/unskilled men required. For this schedule. The day to day manpower requirements arc given below the network.

For allocating the limited resources day-by-day scheduling heuristics can be applied.

Day 1 : Activity which can start on this day is A.

S.M.R (skilled men required) are 7 and

U.M.R (unskilled men required) arc 12.

Schedule activity A.

Day 2 : Activities B, C, D and E arc candidates for the resources, and no activity is already in progress.

Priority order of these activities according to slack is C, D, B and E.

Schedule C, slack = 0, S.M.R. = 4, U.M.X. = 9.

Scheme D, slack = 0, S.M.R. = 3, U.M.K. = 6.

Schedule B, stack = 3, S.M.R. = 1, U.M.R. = 3.

Activity E, slack = 4, S.M.R. = 5, postpone to next day.

Day 3 : Same as for day 2.

Day 4 : Activities C and B are already in progress.

Continue C. slack = 0. S.M.R. = 4, U.M.R. = 9.

Continue B. slack = 3. S.M.R. = 1, U.M.R. = 3.

Activity H. slack = 0, S.M.R. = 6 and U.M.R. = 10 is a critical activity. Its stan is possible only if the non-critical activity B is rescheduled.

Day 9 : On this day activities F and I are candidates for the resources. I is critical, while F has a slack of one day. Though, skilled men required by Fand I are 10, the limit on unskilled men is exceeded. Therefore, schedule activity I and postpone F.

Example 6:

A source memory has six characters with the following probabilities of transmission:

A	*B*	*C*	*D*	*E*	*F*
1/3	*1/4*	*1/8*	*1/8*	*1/12*	*1/12*

Divide the Shannon-Fano encoding procedure to obtain uniquely decodable code to the above message ensemble. What is the average length, efficiency and redundancy of the code that you obtain.

Solution:

The ensembles messages are already in descending order of probabilities.

Divide the elements of set X into subsets, X_1 and X_2 with approximately equal probabilities as:

$$X_1 = \{A, B\} \text{ and } X_2 = \{C, D, E, F\}$$

$$P(X_1) = 1/3 + 1/4 = 7/12$$

and P (X2) = 1/8 + 1/8 + 1/12 + 1/2 = 5/12.

Further divide the set X_2 into two subsets of equal probabilities as shown below:

Subsets	*Probability*
x_{21} = {C, D} = {1/8, 1/8}	$p(X_{21}) = 1/4$
X_{22} = {E, F} = {1/12, 1/12}	$P(X_{22}) = 1/16$

Assign binary number 0 and 1 to first position of all code words in X_1 and X_2, respectively as shown below:

Character	***Probabilities***	***Partitioning (p)***	***Code Word***	***Code Word Length (l)***
A	1/3	X_{11}	00	2
B X_1	1/4	X_{12}	10	2
C	1/8	X_{211}	100	3
D X_2	1/8 X_{21}	X_{212}	101	3
E	1/12	X_{221}	110	3
F	1/12 X_{22}	X_{222}	111	3

Subsets X_{21} and X_{22} contain two elements each, therefore these can be further subdivided into two subsets as shown below:

Subsets	*Probability*
X_{211} = {C} = {1/8}	$P(x_{211}) = 1/8$
X_{212} = {D} = {1/8}	$P(x_{212} = 1/8$
X_{221} = {E} = {1/12}	$P(x_{221}) = 1/12$
X_{222} = {F} = {1/12}	$P(x_{222}) = 1/12$

Assignment of binary number to these subsets is already shown above

(a) The entropy of the source is given by:

$$H(X) = -\sum_{i=1}^{6} P\{x_1\} \log P\{x_i\}$$

$$= -\left[\frac{1}{3}\log\frac{1}{3} + \frac{1}{4}\log\frac{1}{4} + \frac{2}{8}\log\frac{1}{8} + \frac{2}{12}\log\frac{1}{12}\right]$$

$$= \frac{1}{3}\log 3 + \frac{1}{4}\log 4 + \frac{2}{8}\log 8 + \frac{2}{12}\log 12$$

$$= \frac{1}{3}\log 3 + \frac{1}{4}\log(2)^2 + \frac{2}{8}\log(2)^3 + \frac{2}{12}\log(2^2 \times 3)$$

$$= \left(\frac{1}{2} + \frac{4}{12} + \frac{6}{8}\right)\log 2 + \left(\frac{1}{3} + \frac{2}{12}\right)\log 3$$

$$= \frac{19}{12} + \frac{1}{2}\log 3 = 2.3752 \text{ bits.}$$

(b) Average code length of the message is given by:

$$L = \sum_{i=1}^{6} l_i p(x_i) = \frac{2}{3} + \frac{2}{4} + \frac{3}{8} + \frac{3}{8} + \frac{3}{12} + \frac{3}{12} = \frac{29}{12} \text{ bits per symbol}$$

(c) Efficiency of the code:

$$\eta = \frac{H(X)}{L} = \frac{2.3752}{29/12} = \frac{12 \times 2.3752}{29} = 0.9828$$

(d) Redundancy of the code, b = 1 – h = 0.0172.

Example 7:

Find the capacity of the memoryless channel specified by the channel matrix

$$P = \begin{bmatrix} 1/2 & 1/4 & 1/4 & 0 \\ 1/4 & 1/4 & 1/4 & 1/4 \\ 0 & 0 & 1 & 0 \\ 1/2 & 0 & 0 & 1/2 \end{bmatrix}$$

Solution:

The capacity of the memoryless channel is given by

$$C = \text{Max } I(X, Y) = \text{Max } \{H(X) + H(Y) - H(X, Y)\}$$

$$= -\sum_{i=1}^{4} P\{x_i, y_j\} \log P\{x_i, y_j\};\ j = 1, 2, 3, 4$$

where $P(x_i, y_1) = (1/2, 1/4, 1/4, 0)$; $P(x_i, y_2) = (1/4, 1/4, 1/4, 1/4)$

$P(x_i, y_3) = (0, 0, 1, 0)$; $P(x_i, y_4) = (1/2, 0, 0, 1/2)$

Thus $C = \frac{1}{2}\log\frac{1}{2} + 2\left(\frac{1}{4}\log\frac{1}{4}\right) + 4\left(\frac{1}{4}\log\frac{1}{4}\right) + 1\log 1 + 2\left(\frac{1}{2}\log\frac{1}{2}\right)$

$$= \frac{3}{2}\log 2 + 3\log 2 = \frac{9}{2} \text{ bits per symbol.}$$

Example 8:

There are 12 coins, all of equal weight except one which may be lighter or heavier. Using concepts of information theory show that it is possible to determine which coin in heavier.

Solution:

There are 12 coins and one of which is heavier (identical in appearance to all others). In order to isolate the heavier coin an equal arm balance is used for weighing. Here weighing implies putting of a subset of the coins on each of the balance pans and then observing the result. The problem is to find the heavy coin in the smallest number of weighings.

Since we have 12 coins, therefore minimum three weighings are required to isolate the heavy coin. The procedure of isolating the heavier coin can be summarized as follows: place four coins on each pan of the balance. Two possibilities may arise: (a) if left (or right) pan is heavier, then the heavy coin is on the left (or right pans); (b) if pan are balanced, then the heavy coin is among the four not weighed. In the case of (a), the heavy coin is found in two weighings whereas in case of (b) one more weighing suffices.

The expected amount of information necessary to isolate the heavy coin in case there are n coins is given by

$$H\ \{(1/n), (1/n), (1/n), \ldots ,(1/n)\} = -\log_2 (1/n) = \log_2 n \text{ bits}$$

The all possible cases would be

Coin 1 is heavy with probability (1/n)

Coin 2 is heavy with probability (1/n)

... ...

Coin n is heavy with probability (1/n)

Thus, in other words, $\log_2 n$ bits of information has to be accumulated to isolate the heavy coin. Hence, for n = 12, the expected amount of information received is $\log_2 12$ bits.

Suppose in the first weighing, there are x coins on each pan and 12 – 2x coins not weighed. Since coins are equally likely to be the odd one, therefore at each weighing, the following three probabilities may arise

(i) P {left pan down} $= \frac{x}{12}$

(ii) P {right pan down} $= \frac{x}{12}$

(iii) P {pans balanced} $= \frac{12-2x}{12}$

Then the expected amount of information obtained by the outcome of the weighting is given by

$$H = \left\{\frac{x}{12}, \frac{x}{12}, \frac{(12-2x)}{12}\right\}$$

If the total number of coins are divisible by 3, then x = 12/3 = 4, and expected amount of information necessary to isolate the heavy coin is

$$H\ (1/3,\ 1/3,\ 1/3) = -\log_2\ (1/3) = \log_2 3$$

In general, if one is able to get maximum $\log_2 3$ bits of information in each weighing, then k number of weighings provide k $\log_2(3)$ bits of information. To isolate the heavy coin after these k number of weighings, it is required that k $\log_2 3 \geq \log_2$ n or $3^k \geq n$.

EXERCISES

1. An alphabet consists of 8 consonants and 8 vowels. Suppose that all the letters of the alphabet are equally probable and that there is no inter symbol influence. If consonants are always understood correctly, but vowels are understood correctly only half the time, being mistaken for other vowels the other half of the time, all vowels being involved in errors the same percentage of the time, what is the average rate of information transmission.
2. Evaluate the entropy associated with the following probability distribution.

Event :	A	B	C	D
Probability :	1/2	1/4	1/8	1/8

3. Prove that $H(p_1\ p_2\ ...,\ p_n)$ £ $\log_2$ n, and equality holds if and only if,

$$p_k = 1/n;\ k = I,\ 1,\ ...,\ n.$$

4. If H denotes the entropy function, then prove that

$$H\ (p_1, p_2, ..., p_n, q_1, q_2, ..., q_m) = H\ (p_1, p_2, ..., p_n) + p_n$$

$$H\left(\frac{q_1}{p_n}, \frac{q_2}{p_n}, \ldots, \frac{q_m}{p_n}\right)$$

where $p_n = q_1 + q_2 + \ldots + q_m$. Verify the formula, defining additivity of entropies for events A, B, and C with probabilities 1/5, 4/15 and 8/15 respectively.

5. If H denotes the entropy function, then prove that:

   $H(p_1, p_2, \ldots, p_{n-1}, q_1, q_2, \ldots, q_m) = H(p_1, p_2, \ldots, p_n) + p_n$

   $$H\left(\frac{q_1}{p_n}, \frac{q_2}{p_n}, \ldots, \frac{q_m}{p_n}\right)$$

   where p_n $p_n = \sum_{k=1}^{m} q_k$

6. Verify the rule of the additivity of entropies for events A, B, C with probabilities 1/5, 4/15 and 8/15 respectively.

7. A word consists of three letters with respective probabilities 5/12, 1/2 and 1/12. Find the average amount of information associated with the transmission of letters.

8. Write a critical essay on information theory emphasizing the basic concepts.

9. Define entropy function and establish its formal requirements.

10. Show that the entropy function is maximum when mutually exclusive events are equi-probable. Show that partitioning of events into sub-events cannot decrease the entropy of the system.

11. The following two finite probability schemes are given by $(p_1, p_2, \ldots, p_n)$, and $(q_1, (q_2, \ldots, q_n)$, with

    $$\sum_{i=1}^{n} p_1 = \sum_{i=1}^{n} q_i$$

    Then show that $-\sum_{i=1}^{n} p_1 \log p_i \leq -\sum_{i=1}^{n} q_i \log q_i$

    with equality if and only if $p_i = q_i$ for all i.

12. Let X be a discrete random variable taking values $x_1, x_2, \ldots, x_n$ with probability $p(X = x_k) = p_k$, $k = 1,2, \ldots, n$ where $p_k \geq 0$ and p, $p_1 + p_2 \ldots + p_n = 1$. Define the entropy $H(p_1, p_2, \ldots, p_n)$ of the probability distribution to X and prove that

$$H(p_1, p_2, ..., p_n) = H(p_1, p_2, ..., p_{n-1}, p_n) + (p_{n-1}, + p_n)$$
$$H\left(\frac{p_n - 1}{p_n + p_{n-1}}, \frac{p_n}{p_n + p_n - 1}\right)$$

13. In a certain community 25 per cent of all girls are blondes and 75 per cent of all blondes have blue eyes. Also 50 per cent of all girls in the community have blue eyes. If you know that a girl has blue eyes, how much additional information do you get by being informed that she is blonde ?

14. Apply Shannon-Fano encoding procedure to the following message ensemble:

[X] :	x_1	x_2	x_3	x_4	x_5	x_6	x_7	x_8	x_9
[P] :	0.49	0.14	0.14	0.07	0.07	0.04	0.02	0.02	0.01

15. Find the capacity of the memoryless channel specified by the channel matrix:

$$p = \begin{bmatrix} 1/3 & 2/3 & 0 \\ 2/3 & 1/3 & 0 \\ 0 & 0 & 0 \end{bmatrix}.$$

16. A transmitter and receiver has an alphabet consisting of three letters each. The joint probabilities for communication are given below:

$P(x_i, y_j)$	y_1	y_2	y_3
x_1	0.45	0.45	0.01
x_2	0.02	0.02	0.01
x_3	0.01	0.02	0.01

Determine the different entropies for this channel.

17. Apply Shannon's encoding procedure to the following message ensemble.

[X]	:	A	B	C	D
[P]	:	0.4	0.3	0.2	0.1

18. The activities comprising a certain project have been identified as follows:

Activity	*Preceding Activity*	*Duration (Weeks)*	*No. of Men Required*
A	–	4	1
B	–	7	1
C	–	8	2
D	A	5	3
E	C	4	1
F	B, E	4	2
G	C	11	2
H	G, F	4	1

(a) For the above project draw the network, determine the critical path and its duration.

(b) If there were only three men available at any one time, how long would the project take and how would you allocate the men to the activities?

(c) If there were no restrictions on the amount of labour available, explain how you might schedule the activities commencing on different criteria that might be used.

19. A simple project has the following time and resource data :

Activity	*Preceding Activity*	*Duration (Days)*	*Labour Requirement*
A	–	1	2
B	–	2	1
C	A	1	1
D	–	5	1
E	B	1	1
F	C	1	1

(a) Determine the minimum project schedule.

(b) Find the project schedule if only 2 men are available.

(c) Determine the number of men required if the project duration in (a) cannot be extended.

20. For a project consisting of several activities, the duration and required resources for carrying out each of the activity and their availabilities are given below.

Activity		*Resources Required*	
i-j	*Equipment*	*Operators*	*Duration (Days)*
1-2	X	30	4
1-3	Y	20	3
1-4	Z	20	6
2-4	X	30	4
2-5	Z	20	8
3-4	Y	20	4
3-5	Y	20	4
4-5	X	30	6

Resource availability :

Equipment X = 1, equipment Y = 1,

equipment Z = 1, no. of persons = 50.

(a) Draw the network, identify critical path and compute the total float for each of the activities.

(b) Find the project completion time under the given resource constraints.

Time	:	0	1 to 8	9 to 12	13 to 18	19 to 21
Operators	:	0	50	40	50	20.

2

Dynamic and Linear Programming

INTRODUCTION

The dynamic programming problem can be decomposed or divided into a sequence of smaller sub-problems called *stages*. At each stage there are a number of decision alternatives (courses of action) and a decision is made by selecting the most suitable alternative. Stages very often represent different time periods in the planning period of the problem, places, people or other entities. For example, in the replacement probably, each year is a stage, in the salesman allocation problem, each territory represents a stage.

Each stage in a dynamic programming problem has a certain number of states associated with it. These states represent various conditions of the decision process at a stage. The variables which specify the condition of the decision process or describe the status of the system at a particular stage are called *state variables*. These variables provide information for analysing the possible effects that the current decision, could have upon future courses of action. At any stage of the decision-making process there could be a finite or infinite number of states. For example, a specific city is referred to as state variable in any stage of the shortest route problem.

Return function : At each stage, a decision is made which can affect the state of the system at the next stage and help in arriving at the optimal solution at the current stage. Every decision that is made has its own merit in terms of worth or benefit associated with it and can be described in an algebraic equation form. This equation is generally called a return function, since for every set of decisions, a return on each decision is obtained.

This *return function* in general depends on the *state variable* as well as the decision made at a particular stage. An *optimal policy* or *decision* at a stage yields optimal (maximum or minimum) return for a given value of the state variable. It can be seen that at each stage of the problem, there are two inputs: state (variable) s_n and decision (variable) d_n. The state (variable) is the state input which relates the present stage back to the previous stage.

For example, the current state s_n, provides complete information about various possible conditions in which the problem is to be solved when there are n stages to go. The decision d_n is made at stage n for optimizing the total return over the remaining n–1 stages. The decision d_n which optimizes the output at stage n produces two outputs: (i) the return function $r_n(s_n, d_n)$ and (ii) the new state variable s_{n-1}. The return function which is expressed as function of the state variable, s_n and the decision (variable), d_n indicates about the state of the process at the beginning of the next stage (stage n–1), and is denoted by *transition function* (state transformation)

$$s_{n-1} = t_n(s_n, d_n),$$

where t_n represents a state transformation function and its form depends on the particular problem to be solved. This formula allows to go from one stage to another.

DEVELOPING OPTIMAL DECISION POLICY

As pointed out earlier, dynamic programming is an approach in which the problem is broken down into a number of smaller sub-problems called *stages*. These sub-problems are then solved sequentially until the original problem is finally solved. A particular sequence of alternatives (courses of action) adopted by the decision-maker in a multi-stage decision problem is called a policy. The optimal policy, therefore, is the sequence of alternatives that achieves the decision-maker's objective. The solution of a dynamic programming problem is based upon *Bellman's principle of optimality* (recursive optimization technique) which states:

The optimal policy must be one such that, regardless of how a particular state is reached, all later decisions (choices) proceeding from that state must be optimal.

Based on this principle of optimality, we find the best policy by solving one stage at a time, and then sequentially adding a series of one-stage-problems that are solved until the overall optimum of the initial problem is obtained. The solution procedure is based on a *backward induction process and forward induction process*. In the first process, the problem is solved by solving the problem in the last stage and working backwards towards the first stage, making optimal decisions at each stage of the problem. In certain cases, the second process is used to solve a problem by first solving the initial stage of the problem and working towards the last stage, making an optimal decision at each stage of the problem.

It was also mentioned earlier that the exact recursion relationship would vary according to the nature of the problem to be solved by dynamic programming. It is clear from the above discussion that one stage return is given by

$$f_1 = r_1 (s_1, d_1)$$

and the optimal value of f_1 under the state variable s_1 can be obtained by selecting a suitable decision variable d_1. That is,

$$f_1^*(s_1) = \underset{d_1}{\text{Out}} \{r_1(s_1, d_1)\}.$$

The range of d_1 is determined by s_1, but s_1 is determined by what has happened in Stage 2. Then in Stage 2 the return function will take the form:

$$f_1^* (s_2) = \underset{d_1}{\text{Out}} \left\{r_2 (s_2)* f_1^* (s_1)\right\};\ s_1 = t_2 (s_2, d_2)$$

By continuing the above logic recursively for a general n stage problem, we have

$$f_2^* (s_n) = \underset{d_n}{\text{Out}} \left\{r_n (s_n, d_n)* f_2^* (s_{n-1})\right\};\ s_{n-1} = t_n (s_n, d_n).$$

Here the symbol * denotes any mathematical relationship between s_n and d_n, including addition, subtraction, multiplication, etc.

The General Algorithm

The procedure for solving a problem by using the dynamic programming approach can be summarized in the following steps:

Step 1. Identify the problem decision variables and specify objective function to be optimized under certain limitations, if any.

Step 2. Decompose (or divide) the given problem into a number of smaller sub-problems (or stages). Identify the state variables at each stage and write down the transformation function as a function of the state variable and decision variable at the next stage.

Step 3. Write down a general recursive relationship for computing the optimal policy. Decide whether to follow the forward or the backward method to solve the problem.

Step 4. Construct appropriate tables to show the required values of the return function at each stage as shown in Table 1.

Step 5. Determine the overall optimal policy or decisions and its value at each stage. There may be more than one such optimal policy.

Table 1 : Stage 1

	Decision, d_n →	$\frac{f_n(s_n, d_n)}{d_n}$	*Optimal Return*	*Optimal*
Decision				
States, s_n ↓			$f_n^*(s_n)$	d_n^*

DYNAMIC PROGRAMMING UNDER CERTAINTY

Dynamic programming under certainty involves problems where the problem conditions at each stage, *i.e.,* state variables are known with certainty. In deterministic dynamic programming there is neither uncertainty nor probability distribution associated with the state in various stages of the decision process.

Model I. Shortest Route Problem

It is illustrated by some example

Model II. Single Additive Constraint, Multiplicative Separable Return

The general form of the recursive equation to solve such a problem by dynamic programming is illustrated by considering the following problem. Suppose we have separable return functions $f_j(d_j)$, j = 1, 2, ... ,n and desire to

$$\text{Maximize } Z = \{f_1(d_1).\ f_2(d_2),\ ...\ ,f_n(d_n)\}$$

subject to the constraints

$$a_1d_1 + a_2d_2 + ... + a_nd_n = b$$

and $$d_j,\ a_j,\ b \geq 0 \text{ for all } j = 1, 2, ... ,n$$

where j = jth number of stage ($j = 1, 2, \ldots, n$)

d_j = decision variable at jth stage; a_j = constant.

Let us now define state variables, $s_1, s_2, \ldots, s_n$ such that

$$s_n = a_1d_1 + a_2d_2 + \ldots + a_nd_n = b$$

$$s_{n-1} = a_1d_1 + a_2d_2 + \ldots + a_{n-1}\,d_{n-1} = s_n - a_nd_n$$

$$\vdots$$

$$s_{j-1} = s_j - a_jd_j$$

$$\vdots$$

$$s_1 = s_2 - a_2d_2$$

In general, it may be noted that the state transition function takes the form

$$s_{j-1} = t_j\,(s_j, d_j);\ j = 1, , \ldots ,n,$$

i.e., a function of next state and decision variables.

At the nth stage, sn is expressed as the function of the decision variables. Thus the maximum value of Z denoted by $f_n{}^*\,(s_n)$ for any feasible value of sn is given by

$$f_n{}^*\,(s_n) = \underset{d_j > 0}{\text{Maximum}}\ \{f_1\,(d_1)\ ;\ f_2\,(d_2) \ldots f_n\,(d_n)$$

subject to the constraint $s_n = b$

Now for a moment holding a particular value of d_n fixed, the maximum value of Z will be given by

$$f_n\,(d_n) * \underset{d_j > 0}{\text{Max}}\ \{f_1\,(d_1)\ .\ f_2\,(d_2) \ldots f_{n-1}\,(d_{n-1})\};\ j = 1, 2, \ldots ,n-1$$

$$= f_n\,(d_n) * f_{n-1}{}^*\,(s_{n-1})$$

The maximum value $f_{n-1}{}^*\,(d_{n-1})$ of Z due to decision variables d_j ($j = 1, 2, \ldots , n-1$) depends upon the state variable $s_{n-1} = t_n\,(s_n, d_n)$. The maximum of Z for any feasible value of all decision variables will be given by

$$f_j{}^*\,(s_j) = \underset{d_j > 0}{\text{Max}}\ [f_j\,(d_j) * f_{j-1}{}^*\,(s_j - 1)];\ j = n, n-1, \ldots ,2$$

$$f_1\,(s_1) = f_1\,(d_1)$$

where $s_{j-1} = t_j\,(s_j, d_j)$

The value of $f_j{}^*\,(s_j)$ represents the general recursive equation.

Model III. Single Additive Constraint, Additive Separable Return

Consider the problem

$$\text{Minimize } Z = [f_1 (d_1) + f_2 (d_2) + \dots + f_n (d_n)]$$

subject to the constraint

$$a_1d_1 + a_2d_2 + \dots + a_nd_n \geq b$$

and a_j, d_j, $b \geq 0$ for all j

Proceed in the same manner as in Model II. Defining state variables $s_1, s_2, \dots s_n$ such that

$$s_n = a_1d_1 + a_2d_2 + a_nd_n \geq b$$

$$s_{n-1} = s_n - a_nd_n$$

$$\vdots$$

$$s_{j-1} = s_j - a_jd_j;\ j = 1, 2, 3, \dots, n$$

Let $f_n^* (s_n) = \underset{d_j > 0}{\text{Min}} \sum_{n=1}^{n} f_j(d_j)$

such that $s_n \geq b$

The general recursive equation for obtaining the minimum value of Z for all decision variables and for any feasible value of all decision variables is given by

$$f_j^* (s_j) = \underset{d_j > 0}{\text{Min}} [f_j (d_j) + f_{j-1}^* (s_{j-1})].\ j = 2, 3, \dots, n$$

$$f_1^* (s_1) = f_1 (d_1)$$

where $s_{j-1} = t_j (s_j, d_j)$

Mode IV. Single Multiplicative Constraint, Additively Separable Return

Consider the problem

$$\text{Minimize } Z = [f_1 (d_1) + f_2 (d_2) + \dots + f_n (d_n)]$$

subject to the constraint

$$d_1 . d_2 . d_3. \dots . d_n \geq b$$

and d_j, $b \geq 0$ for all j

Proceed in the same manner as in Model II. Define state variables $s_1, s_2, \dots, s_n$ such that

$$s_n = d_n \cdot d_{n-1} \ldots d_2 \cdot d_1 \geq b$$

$$s_{n-1} = d_{n-1} \cdot d_n - 2 \ldots d_2 \cdot d_1 = s_n/d_n$$

$$\vdots$$

$$s_{j-1} = s_j/d_j;\ j = 2, 3, \ldots, n$$

Let $f_n^* (s_n) = \underset{d_j > 0}{\text{Min}} \sum_{j=1}^{n} f_j(d_j)$ such that $s_n \geq b$

The general recursive equation for obtaining the minimum value of Z for all decision variables and for any feasible value of decision variable will be given by

$$f_j^* (s_j) = \underset{d_j > 0}{\text{Min}} [f_j (d_j) + f_{j-1}^* (s_{n-1})];\ j = 2, 3, \ldots, n$$

$$f_j^* (s_1) = f_1 (d_1)$$

and $s_{j-1} = t_j (s_j, d_j)$

INCONSISTENCY AND REDUNDANCY IN CONSTRAINT EQUATIONS

Redundancy is constraint equations.

By redundancy in constraint equations we mean that the system has more than enough number of constraint equations, in other words it has more constraint equations than the number of variables.

This is the situation when

$$r (A) = r (A\ b) = k \leq n < m.$$

In this case there will be $(m - k)$ redundant equations.

The Inconsistency

As already defined, the set of constraints (linear equations) is said to be inconsistent if

$$r (A) \neq r (Ab).$$

Before solving a LP. problem by simplex method, we should have $r (A) = r (Ab)$ *i.e.,* the constraint aquariums (after introducing the slack and artificial variables) shows be consistent. Since in simplex method we always have

$$r (A) = r (A \times b) = m.$$

If the system A x = b involves artificial variables, then we cannot say whether this system is consistent or there is any redundancy. Below we give the cases to decide about the consistency and redundancy in such systems.

Case I. If at least one artificial vector appears in the basis B at a positive level *i.e.*, the value of at least one artificial variables corresponding to artificial vector in B is non-zero and the optimality condition is satisfied (at any iteration), then there exists no feasible solution of the problem.

Case II. If the basis B contains no artificial vector and the optimality condition is satisfied (at any iteration) then the current solution is a B.F.S. of the problem.

Case III. If one or more artificial vector appears in the basis B at a zero level *i.e.*, the value of the artificial variables corresponding to artificial vectors in B are zero and the optimality condition is satisfied (at any iteration) then the system is consistent. Furthermore if $y_{ij} = 0 \; \forall$ j and $x_{Br} = 0$ and r corresponds to the row containing an artificial vector, then the rth constraint equation is redundant.

TO DETERMINE IMPROVED B.F.S. FROM A GIVEN B.F.S.

Theorem:

Let $x_B = B^{-1} b$ be a B.F.S. of a linear programming problem with $Z = c_B x_B$ as the value of the objective function. If for any column aj in A, but not in B, the condition $c_j - Z_j > 0$ or $Z_j - c_j < 0$ holds and if at least one $y_{ij} > 0$, $i = 1, 2, \ldots, m$, then it is possible to obtain a new B.F.S. by replacing one of the columns in B by α_j and if Z' is the new value of objective function then $Z' \geq Z$.

If the initial B.F.S. $x_B = B^{-1} b$ is non-degenerate then $Z' > Z$.

Proof:

Consider the L.P.P.

$$\text{Max. } Z = c\,x, \text{ s.t. } A\,x = b, \; x \geq 0,$$

where $A = (\alpha_1, \alpha_2, \ldots, \alpha_N)$, $N = m + n$,

Basis matrix $B = (\beta_1, \beta_2, \ldots, \beta_m)$.

Let x_B $(x_{B1}, x_{B2}, \ldots, x_{Bm})$ be a B.F.S. of the given L.P.P.

Since the vectors $\beta_1, \beta_2, \ldots, \beta_m$ are in the basis of A, therefore we can express α_j as a linear combination of b's.

$$\therefore \quad \alpha_j = \sum_{i=1}^{m} y_{ij}\beta_i = y_{1j}\,\beta_1 + y_{2j}\,\beta_2 + \ldots + y_{mj}\,\beta_m. \qquad \ldots(1)$$

If $y_{rj} \neq 0$ then α_j can replace br in B and B is still a basis matrix.

Let $y_{rj} \neq 0$ then from (1), we have

$$\Rightarrow \quad \beta_r = \frac{1}{y_{rj}}\alpha_j - \frac{y_{1j}}{y_{rj}}\beta_1 - \ldots - \frac{y_{(r+1)j}}{y_{rj}}\beta_{r+1} - \ldots - \frac{y_{mj}}{y_{rj}}\beta_m$$

$$\Rightarrow \quad \beta_r = \frac{1}{y_{rj}}\alpha_j - \sum_{i=1}^{m}\frac{y_{ij}}{y_{rj}}\beta_i. \qquad \ldots(2)$$

Now, we have

$$\Rightarrow \quad Bx_B = b$$

$$\Rightarrow \quad (\beta_1, \beta_2, \ldots, \beta_r, \ldots, \beta_m)\,(x_{B1}, x_{B2}, \ldots, x_{Br}, \ldots, x_{Bm}) = b$$

$$\Rightarrow \quad \beta_1\,x_{B1} + \beta_2\,x_{B2} + \ldots + \beta_r\,x_{Br} + \ldots + \beta_m\,x_{Bm} = b$$

$$\Rightarrow \quad \sum_{i=1}^{m}\beta_i x_{Bi} + \beta_r x_{Br} = b \qquad \ldots(3)$$

Putting for br from (2) in (3), we get

$$\Rightarrow \quad \sum_{i=1}^{m}\beta_i\left[x_{Bi} - \frac{y_{ij}}{y_{rj}}x_{Br}\right] + \frac{x_{Br}}{y_{rj}}\alpha_j = b \qquad \ldots(4)$$

$$\Rightarrow \quad \sum_{i=1}^{m}x_{Bi}'\beta_i + x_{Br}'\alpha j = b, \qquad \ldots(5)$$

where $x_{Bi}' = x\!-\! - x_{Br}\dfrac{y_{ij}}{y_{rj}}$, $i = 1, 2, \ldots, m,\ i \neq r$

$$x_{Br}' = \frac{x_{Br}}{y_{rj}},\ i = r$$

Comparing (3) and (5), we observe that new basic solution of A x = b is given by

$$x'_B = (x'_{Bi}, x'_{Br}),\ i = 1, 2, \ldots, m,\ i \neq r$$

where the values of x'_{Bi}, x'_{Br} are given by (6).

This basic solution will be feasible if we have

$$x_{Bi} - \frac{y_{ij}}{y_{rj}} x_{Br} \geq 0,\ i = 1, 2, \ldots m,\ i \neq r$$

and $\dfrac{x_{Br}}{y_{rj}} \geq 0,\ i = r.$ ...(7)

Since x_B is the initial B.F.S. we have

$$x_{Br} \geq 0,\ r = 1, 2, \ldots ,m.$$

Thus we observe that (7) will hold only if

$$y_{rj} > 0 \text{ and } y_{ij} \leq 0,\ i = 1, 2, \ldots ,m,\ i \neq r.$$

If $y_{rj} > 0$ and $y_{ij} > 0$, then (7) is satisfied only if

$$\Rightarrow \quad \frac{x_{Bi}}{y_{ij}} - \frac{x_{Br}}{y_{rj}} \geq 0$$

$$\Rightarrow \quad \frac{x_{Br}}{y_{rj}} \leq \frac{x_{Bi}}{y_{ij}}$$

$$\Rightarrow \quad \frac{x_{Br}}{y_{rj}} = \min_i \left\{ \frac{x_{Bi}}{y_{ij}}, y_{ij} > 0 \right\} = v. \quad \text{...(8)}$$

Thus, a new B.F.S. can be obtained from the initial B.F.S. by removing the column vector br of the basis matrix B by a_j if r is to be selected such that

$$v = \frac{x_{Br}}{y_{rj}} = \min_i \left\{ \frac{x_{Bi}}{y_{ij}}, y_{ij} < 0 \right\}.$$

If we have v = 0, which is possible only when $x_{Br} = 0$ then it means that the initial B.F.S. is degenerate.

Now we shall prove $Z' \geq Z$.

The value of the objective function for the initial B.F.S. x_B is

$$Z = c_B\, x_B$$

$$= (c_{B1}, c_{B2}, \ldots ,c_{Bm})\ (x_{B1}, x_{B2}, \ldots ,x_{Bm})$$

$$= \sum_{i=1}^{m} c_{Bi} x_{Bi}$$

Corresponding to the new B.F.S. x'_B, the value of the objective function is Z', so we have

$$Z' = \sum_{i=1}^{m} c'_{Bi}\, x'_{Bi}, \qquad ...(9)$$

where c_{Bi}' are the coefficients of the basic variables x_{Bi}' $(i = 1, 2, \ldots, m)$ in the objective function.

Obviously, $c_{Bi}' = c_{Bi}$, $i = 1, 2, \ldots, m$, $i \neq r$

and $c_{Br}' = c_j$.

$\therefore$ we can write

$$Z' = \sum_{i=1}^{m} c_{Bi}\, x'_{Bi} + c_j x'_{Br}. \qquad ...(10)$$

Substituting for x_{Bi}', x_{Br}' from (6) in (10), we get

$$Z' = \sum_{i=1}^{m} c_{Bi}\left(x_{Bi} - x_{Br}\frac{y_{ij}}{y_{rj}}\right) + c_j\frac{x_{Br}}{y_{rj}}$$

$$= \sum_{i=1}^{m} c_{Bi}\left(x_{Bi} - x_{Br}\frac{y_{ij}}{y_{rj}}\right) + c_j\frac{x_{Br}}{y_{rj}}$$

Since the term $c_{Bi}\left(x_{Bi} - x_{Br}\frac{y_{ij}}{y_{rj}}\right) = 0$ when $i = r$ therefore it can be included in the summation without changing the value of Z'.

$$\therefore \quad Z' = \sum_{i=1}^{m} c_{Bi} x_{Bi} - \frac{x_{Br}}{y_{rj}}\sum_{i=1}^{m} c_{Bi} y_{ij} + \frac{x_{Br}}{y_{rj}} c_j$$

$$= Z + \frac{x_{Br}}{y_{rj}}(c_j - Z_j), \text{ where } Z_j = \sum_{i=1}^{m} c_{Bi} y_{ij}$$

$$= Z + v\,(c_j - Z_j), \text{ from (8).} \qquad ...(11)$$

From (11), we observe that the new value of the objective function is equal to the original value of the objective function plus the quantity $v\,(c_j - Z_j)$.

We have $Z' \geq Z$ if $v\,(c_j - Z_j) \geq 0$.

Since $v \geq 0$ therefore $Z' \geq Z$ only if $c_j - Z_j \geq 0$.

Hence by choosing the vector α_j for which $c_j - Z_j > 0$ and at least one $y_{ij} > 0$, we obtain a new improved value of the objective function.

If the initial B.F.S. is non-degenerate then $v > 0$ and in that case $Z' > Z$.

SIMPLEX METHOD BY INVERSE OF A MATRIX

Consider any n × n non-singular real matrix A.

Also consider a system of equation A x = b, x ≥ 0,

where b is a dumbly real matrix of order n × 1.

Introduce the artificial variables $x_a \geq 0$ and a dummy objective function Z with cost zero to variables in x and cost (–1) to each artificial variable. Now find the solution of the following linear programming problem using simplex method

$$\text{Max. } Z = 0.\ x - 1.x_{art}$$

subject to A x = b, x, $x_a \geq 0$.

In the last iteration when the column of A become the columns of unit matrix I, the inverse of matrix A is given by those column vectors of the table which were the column of the initial basis.

UNBOUNDED SOLUTION HAVING LINEAR PROGRAMMING PROBLEM

A linear programming problem has unbounded solution if the feasible region is unbounded such that the value of the objective function can be increased indefinitely. It is, however not necessary that an unbounded feasible region should yield an unbounded value for the objective function.

L.P.P. having unbounded feasible region but bounded optimal solution

COMPUTATIONAL PROCEDURE OF THE TWO PHASE METHOD

Phase I.

Step 1. To convert each of the constraints into equations, subtract a surplus variable and add an artificial variables.

Step 2. Assign zero coefficients to each of the primary variables and surplus variables and the coefficient (–1) to each of the artificial variables in the objective function. As a result the objective function changes to

Z = – (sum of the artificial variables)

Step 3. Solve the problem formed in step 2 by applying the simplex method. If the original problem has a feasible solution then this problem shall have an optimal solution with optimal value of the objective function equal to zero as each of the artificial variables will be equal to zero.

If max $Z < 0$ and at least one artificial variables appears in the optimum basis at a positive level then the given problem does not possess any feasible solution.

If max $Z = 0$ and at least one artificial variable appears in the optimum basis at zero level, we proceed to phase II.

If max $Z = 0$ and no artificial variable appears in the optimum basis then we proceed to phase II.

Phase II.

In phase II start with the (optimal) solution contained in the final simplex table of the phase I. Assign the actual costs to the variables in the objective function and a zero cost to every surplus variable. Eliminate the artificial which are non-basic at the and of the phase-I. Remove c_j row values of the optimum table and replace them by c_j values of the original problem. Now, apply simplex algorithm to the problem contained in the new table to obtain the optimal solution.

Disadvantages of Big-M Method Over Two Phase Method

(i) We can always use Big-M method to check the existence of a feasible solution but its computational procedure may be inconvenient because of the manipulation of the constant M.

Two phase method eliminates the artificial variables in the beginning.

(ii) When we solve a problem on a digital computer we have to assign some numerical value to M which must be larger than the values c_1, c_2, ... present in the objective function. But a computer has only a fixed number of digits.

DYNAMIC PROGRAMMING APPROACH FOR SOLVING LINEAR PROGRAMMING PROBLEM

A linear programming problem in n decision variables and m constraints can be converted into an n stage dynamic programming problem with m states. To illustrate, let us consider a general linear programming problem:

$$\text{Maximize } Z = \sum_{j=1}^{n} c_j x_j$$

subject to the constraints

$$\sum_{j=1}^{n} a_{ij} x_j \le b_i;\ i = 1, 2, \dots, m$$

and $x_j \ge 0;\ j = 1, 2, \dots, n$

To solve an LP problem by using dynamic programming, the value of the decision variable x_j is determined at stage j (j = 1, 2, ... n). The value of xj at several stages can be obtained either by the forward or the backward induction method. The state variables at each stage are the amount of resources available for allocation to the current stage and succeeding stages.

Let $a_{1j}, a_{2j}, \dots, a_{mj}$ be amount (in units) of resources i (i = 1, 2, ... ,m) respectively allocated to an activity c_j at jth stage, and fn $(a_{1j}, a_{2j}, \dots, a_{mj})$ be the optimum value of the objective function of a general LP problem for stages j, j + 1, ... ,n, and for states $a_{1j}, a_{2j}, \dots, a_{mj}$. Thus, the LP problem may be defined by a sequence of functions as:

$$f_n (a_{1j}, a_{2j}, \dots, a_{mj}) = \text{Max} \sum_{j=1}^{n} c_j x_j$$

This maximization is taken over the decision variable x_j such that

$$\sum_{j=1}^{n} a_{ij} x_j \le b_i;\ x_j \ge 0$$

The recursive relations for optimization are

$$f_n (a_{1n}, a_{2n}, \dots, a_{mn}) \text{ or } f_n (b_1, b_2, \dots, b_m)$$

$$= \underset{0 \le a_{in} x_n \le b_i}{\text{Max}} \{c_n x_n + f_{n-1} (b_1 - a_{1n} x_n, b_2 - a_{2n} x_n, \dots, b_m - a_{mn} x_n)\}$$

The maximum value b that a variable x_n can assume is

$$b = \text{Min} \left\{ \frac{b_1}{a_{1n}}, \frac{b_2}{a_{2n}}, \dots, \frac{b_m}{a_{mn}} \right\}$$

because the minimum value satisfies the set of constraints simultaneously.

LINEAR PROGRAMMING

The graphical method used is applicable only to solve two-variable LP problems. Finding and evaluating all basic feasible solutions of a LP problem, say with six variables and five constraints may be a long and difficult task. Thus, an efficient computational procedure is required to solve the general class of LP problem. This chapter contains a rigorous discussion on the mathematical method called the *simplex method* developed by *G.B. Dantzig* in 1947 for solving LP problems. The simplex method is an *iterative procedure* and refers to the idea of moving from one extreme (comer) point to another of the solution space or feasible region (convex polyhedron) that is formed by the constraints and non-negativity conditions of the linear programming problem.

Since the number of extreme points (corners or vertices) of a solution space is finite, the method leads to the optimal extreme point (optimal solution) in a finite number of steps or indicates that there exist an unbounded solution. It was observed in the graphical approach to the solution of L.P.P. that in a given situation the set of constraints determines the feasible region and objective function gives the optimal point, the one that minimizes or maximizes, as the case may be. But the graphical method suffers from a great limitation that it can handle problems involving only two decision variables. Whereas in real world situations, we frequently encounter such problems where more than two variables are involved.

Fundamental theorem of linear programming makes it easy to find an optimal solution because it deals with basic feasible solutions only. But, it is also not easy job to enumerate all the B.F.S. even for small values of m (number of constraints) and n (number of variables). The simplex method is an iterative procedure, for finding, in a systematic manner the optimal solution to a linear programming problem.

CANONICAL AND STANDARD FORM OF LP PROBLEM

Canonical form : If the general formulation of LP problem is expressed as:

$$\text{Maximize } Z = \sum_{j=1}^{n} c_j x_j$$

subject to the constraints

$$\sum_{j=1}^{n} a_{ij} x_j \le b_i; \qquad i = 1, 2, ..., m$$

and $x_j \geq 0$; $j = 1, 2, ..., n$

then it is called the canonical form of LP problem. The characteristics of this form are:

(i) The objective function should be of maximization type. If not, then it should be changed to the same by applying the method discussed before.

(ii) All constraints should be of ≤ type except for the non-negativity conditions. An inequality of ≥ type can be changed to an inequality of ≤ type by multiplying it on both sides by – 1.

(iii) All variables must have non-negative values. If any variable, say x_j, is unrestricted in sign (i.e. positive, negative or zero), then it can be replaced by

$$x_j = x_j' - x_j''$$

where x_j' and x_j'' are both non-negative.

(iv) The right-hand side of each constraint should be positive.

Standard Form : If the general formulation of the LP problem as stated in Chapter 3, is expressed as

$$\text{Maximize } Z = \sum_{j=1}^{n} c_j x_j$$

subject to the constraints

$$\sum_{j=1}^{n} a_{ij} x_j \leq b_i; \qquad i = 1, 2, ..., m$$

and $x_j \geq 0$; $j = 1, 2, ..., n$

then it is called the standard/arm of the LP problem. The characteristics of this form are:

(i) All the constraints should be expressed as equations.

(ii) The right-hand-side of each constraint should be made non-negative. If it is not so, this should be done by multiplying both sides of the resulting constraints by – 1.

(iii) The objective function should be of maximization type.

The standard form can also be written in matrix notation as follows:

$$\text{Maximize } Z = ex \qquad ...(1)$$

subject to the constraints

$$Ax = b \qquad ...(2)$$

and $$x > 0, \qquad ...(3)$$

where $c = (c_1, c_2, ..., c_n)$ is the row vector; $x = (x_1, x_2, ..., x_n)^T$ and $b = (b_1, b_2, ..., b_m)$ are column vectors and A is m × n coefficients matrix of rank m.

The LP problem can also be represented in terms of column vectors, $a_1, a_2, ..., a_n$ of matrix A as follows:

$$\text{Maximize } Z = \sum_{j=1}^{n} c_j x_j$$

subject to the constraints

$$\sum_{j=1}^{n} a_j x_j = b$$

and $$x_j \geq 0; \quad j = 1, 2, ..., n.$$

TO REDUCTION BASIC FEASIBLE SOLUTION

Theorem:

The linear programming problem

$$Max.\ Z = c\,x \text{ subject to } A\,x = b,\ x \geq 0,$$

where $A = (\alpha_1, \alpha_2, ..., \alpha_N)$ is the coefficient matrix of order $m \times N$, $(N = m + n)$, has at least one feasible solution then it has at least one basic feasible solution also.

Proof:

Consider an arbitrary feasible solution of the given linear programming problem

$$x^* = (x_1, x_2, ..., x_N),\ x_i \geq 0. \qquad ...(1)$$

Let us assume that k, $(k \leq N)$ variables in x^* have positive values and the remaining N – k variables are zero. We can also assume that the variables have numbered in such a way that the first k variables are non-zero.

Thus $x^* = (x_1, x_2, ..., x_k, 0, 0, ..., 0)$.

Also, we have $\sum_{i=1}^{k} x_i \alpha_i = b$. ...(2)

Now, two possibilities may arise:

The vectors $\alpha_1, \alpha_2, \ldots, \alpha_k$ may be either linearly independent or dependent.

Case I. If $\alpha_1, \alpha_2, \ldots, \alpha_k$ are linearly independent.

If any feasible solution for which the vectors a_i, $i = 1, \ldots, k$ associated with non-zero variables x_i, $i = 1, \ldots, k$ are L.I. is called a basic feasible solution. Thus, x* is a B.F.S. which is also optimal.

Hence, solution x* is degenerate if $k < m$ and non-degenerate if $k = m$.

Case II. If $\alpha_1, \alpha_2, \ldots, \alpha k$ are linearly dependent and $k > m$.

In this case shall reduce the number of non-zero variables step by step until we obtain a B.F.S. from feasible solution.

Since $\alpha_1, \alpha_2, \ldots, \alpha_k$ are linearly dependent therefore there exist scalars $\lambda_1, \lambda_2, \ldots \lambda_k$ such that

$$\lambda_1\, 1 + \lambda_2 + \alpha_2 + \ldots + \lambda_k\, \alpha_k = 0$$

$$\Rightarrow \quad \sum_{o=1}^{k} \lambda_i \alpha_i = 0 \qquad \ldots(3)$$

and at least one $\lambda_1 \neq 0$.

We can assume that at least one λ_i is positive because if none is positive then we can multiply the equation (3) by -1 and get positive λ_i.

Now, suppose

$$v = \max_{1 \le i \le k} \left(\frac{\lambda_i}{x_i} \right). \qquad \ldots(4)$$

Obviously, v is positive since $x_i \geq 0$, $i = 1, 2, \ldots, k$ and at least one λ_i is positive.

Multiplying both sides of (3) by 1/v and subtracting from (2), we get

$$\sum_{i=1}^{k} \left(x_i - \frac{\lambda_i}{v} \right) \alpha_i = b. \qquad \ldots(5)$$

(5) gives $x' = \left[x_1 - \frac{\lambda_1}{v}, x_2 - \frac{\lambda_2}{v}, \ldots, x_k - \frac{\lambda_k}{v}, 0, 0, \ldots, 0 \right]$

a new solution of the matrix equation $A\,x = b$.

Also from (4) for $i = 1, 2, \ldots, k$, we have

$$v \geq \frac{\lambda_i}{x_i} \text{ or } x_i \geq \frac{\lambda_i}{v}$$

or $$x_i - \frac{\lambda_i}{v} \geq 0.$$

Since all the components of x' are non-negative therefore x' is a feasible solution to the given L.P.P.

Also $v = \frac{\lambda_i}{x_i}$, for at least one i

i.e., $v - \frac{\lambda_i}{x_i} = 0$ therefore the new feasible solution x' cannot have more than $k - 1$ non-zero variables.

In this way, we have derived a new feasible solution from the given optimal feasible solution which contains less number of non-zero variables. This solution is B.F.S. if the column vectors associated to non-zero variables in this new solution are L.I. If these associated vectors are not L.I., we shelf repeat the whole reduction procedure as explained above. Continuing in this way for finite number of times we will derive a solution in which columns corresponding to positive variables are L.I. *i.e.,* we will obtain a B.F.S. of the system. Hence the theorem is proved.

THE NOTATIONS OF THE SIMPLEX ALGORITHM

Some notations and definitions which are extremely useful in the discussion of simplex algorithm.

Consider a linear programming problem which after introducing slack and surplus variables is as follows:

$$\text{Max. } Z = c_1 x_1 + c_2 x_2 + \dots + c_n x_n + 0x_{n+1} + 0x_{n+2} + \dots + 0x_{n+m}$$

subject to
$$a_{11} x_1 + a_{12} x_2 + \dots + a_{1j} x_j + \dots + a_{1n} x_n + x_{n+1} = b_1$$
$$a_{21} x_1 + a_{22} x_2 + \dots + a_{2j} x_j + \dots + a_{2n} x_n + x_{n+2} = b_2$$
$$\dots \quad \dots \quad \dots \quad \dots \quad \dots \quad \dots \quad \dots$$
$$\dots \quad \dots \quad \dots \quad \dots \quad \dots \quad \dots \quad \dots$$
$$a_{m1} x_1 + a_{m2} x_2 + \dots + a_{mj} x_j + \dots + a_{mn} x_n + x_{n+m} = b_m,$$

$x_i \geq 0$ for all $i = 1, 2, \dots, N$, where $N = n + m$

$b_1, b_2, \dots, bm$ are all positive.

The above linear programming problem can be easily converted to matrix form

$$\text{Max. } Z = c\,x$$

subject to $A\,x = b,\ x \geq 0$

where $A = [a_{ij}]_{m\times N}$ is the coefficient matrix of order $m \times N$,

$$x = [x_1, x_2, x_3, \ldots, x_n, \ldots, x_N],$$

$$c = [c_1, c_2, \ldots, c_n, 0, 0, \ldots, 0]_{1\times N}$$

and $b = [b_1, b_2, \ldots, b_m]$,

where x and b are column vectors of order $N \times 1$ and $m \times 1$ respectively.

For convenience, we shall represent column vectors by row vectors without using transpose symbol. There should be no confusion in understanding scalar multiplication of two vectors c and x. If we denote the jth column of the matrix A by a_j, $j = 1, 2, \ldots, N$, we can write

$$A = [\alpha_1, \alpha_2, \ldots, \alpha_N].$$

Form a non-singular sub-matrix B of order $m \times m$ whose column vectors are m number of linearly independent columns selected from A. If we denote these columns by $\beta_1, \beta_2, \ldots, \beta_m$ then

$$B = [\beta_1, \beta_2, \ldots, \beta_m].$$

B is called the *basis matrix.*

Since the vectors $\beta_1, \beta_2, \ldots \beta_m$ are linearly independent they form a basis for E^m. Therefore each $\alpha_j \in A \subset E^m$ can be expressed as a linear combination of vectors of B. Thus we can write

$$\alpha_j = \beta_1 y_{1j} + \beta_2 y_{2j} + \ldots + \beta_m y_{mj}$$

$$= (\beta_1, \beta_2, \ldots, \beta_m)\begin{bmatrix} y_{1j} \\ y_{2j} \\ \vdots \\ y_{mj} \end{bmatrix}$$

or $\qquad \alpha_j = BY_j$, where $Y_j = \begin{bmatrix} y_{1j} \\ y_{2j} \\ \vdots \\ y_{mj} \end{bmatrix}_{m\times 1}$

$y_{1j}, y_{2j}, ..., y_{mj}$ are the scalars required to express α_j in such a form.

Also $\qquad \alpha_j = BY \Rightarrow Y_j = B^{-1} a_j$.

The vector Yj will change if the columns of A forming B change.

Any basis matrix B will provide a *basic solution* to A x = b.

The variables corresponding to $\beta_1, \beta_2, ..., \beta_m$ are denoted by $x_{B1}, x_{B2}, ..., x_{Bm}$ respectively and are called *basis variables.*

We denote the column vector of these m basic variables by x_B

$$\therefore \qquad x_B = [x_{B1}, x_{B2}, ..., x_{Bm}]$$

and $$x_B = B^{-1} b.$$

This is called the basic feasible solution (B.F.S.) of the L.P.P.

Since $x_{B1}, x_{B2}, ..., x_{Bm}$ are the basic variables, the remaining n variables belonging to x are called non-basic variables.

We shall denote the coefficients of basic variables $x_{B1}, x_{B2}, ..., x_{Bm}$ in the objective function Z by $c_{B1}, c_{B2}, ..., c_{Bm}$.

Corresponding to any x_B, c_B will represent the row vector contain the constants $c_{B1}, c_{B2}, ..., c_{Bm}$.

$$\therefore \quad c_B = (c_{B1}, c_{B2}, ..., c_{Bm}).$$

Since for any basic feasible solution all non-basic variables are zero therefore the objective function Z becomes

$$Z = c_{B1} x_{B1} + c_{B2} x_{B2} + ... + c_{Bm} x_{Bm} + 0$$
$$= (c_{B1}, c_{B2}, ..., c_{Bm}) (x_{B1}, x_{B2}, ..., x_{Bm})$$
$$= c^B x_B.$$

Finally, we introduce a new variables Z_j , given by

$$Z_j = y_j c_{B1} + y_{2j} c_{B2} + ... + y_{mj} c_{Bm}$$
$$= (c_{B1}, c_{B2}, ... c_{Bm}) (y_{1j}, y_{2j}, ..., y_{mj})$$
$$= c_B y_j.$$

Therefore exists Z_j for each α_j and it changes as the columns of A forming B change.

LINEAR PROGRAMMING FUNDAMENTAL THEOREM

Theorem:

If a linear programming problem

max $Z = cx$ *subject to* $Ax = b,\ x \geq 0,$

where A is an $m \times N$, $(N = m + n)$ *matrix of coefficients given by* $A = (\alpha_1, \alpha_2, \ldots, \alpha_N)$ *has an optimal feasible solution then at least one basic feasible solution must be optimal.*

Proof:

Let $x^* = [x_1, x_2, \ldots, x_N]$

be an optimal feasible solution of the given linear programming problem and

$$Z^* = \sum_{i=1}^{N} c_i x_i$$

be the corresponding optimum value of the objective function.

Also suppose that k ($k \leq N$) variables in x^* are non-zero and the remaining $N - k$ variables are zero. For the sake of convenience we can assume that the first k variables of x^* are non-zero.

Thus $x^* = (x_1, x_2, \ldots, x_k, 0, 0, \ldots, 0)$.

$$\therefore \qquad \sum_{i=1}^{k} x_i \alpha_i = b \qquad \ldots(1)$$

and

$$Z^* = \sum_{i=1}^{k} c_i x_i. \qquad \ldots(2)$$

Now two possibilities may arise:

The vectors $\alpha_1, \alpha_2, \ldots, \alpha_k$ may be either linearly independent or dependent.

Case 1. If $\alpha_1, \alpha_2, \ldots, \alpha_k$ are linearly independent.

Any feasible solution for which the vectors a_i, $i = 1, \ldots, k$ associated with non-zero variables x_i, $i = 1, \ldots, k$ are L.I. is called a basic feasible solution. Thus x^* is a B.F.S. which is also optimal.

Hence the result of the theorem is true.

The solution x is degenerate if $k < m$ and non-degenerate if $k = m$.

Case 2. If $\alpha_1, \alpha_2, \ldots, \alpha_k$ are linearly dependent and $k > m$.

In this case we shall reduce the number of non-zero variables step by step by step until we obtain a B.F.S. from feasible solution.

Since $\alpha_1, \alpha_2, \ldots, \alpha_k$ are linearly dependent therefore there exist scalars $\lambda_1, \alpha_2, \ldots, \lambda_k$ such that

$$\lambda_1 \alpha_1 + \lambda_2 \alpha_2 + \ldots + \lambda_k \alpha_k = 0$$

or $$\sum_{i=1}^{k} \lambda_i \alpha_i = 0 \qquad \ldots(3)$$

and at least one $\lambda_i \neq 0$.

We can assume that at least one li is positive because if none is positive then we can multiply the equation (3) by –1 and get positive li.

Now, suppose $$v = \max_{1 \le i \le k} \left(\frac{\lambda_i}{x_i} \right). \qquad \ldots(4)$$

Obviously v is positive since $x_i \geq 0$, i = 1, 2, ..., k and at least one λ_i is positive.

Multiplying both sides of (3) by 1/v and subtracting from (1), we get

$$\sum_{i=1}^{k} \left(x_i - \frac{\lambda_i}{v} \right) \alpha_i = b. \qquad \ldots(5)$$

(5) gives

$$x' = \left[x_1 - \frac{\lambda_1}{v}, x_2 - \frac{\lambda_2}{v}, \ldots, x_k - \frac{\lambda_k}{v}, 0, 0, \ldots, 0 \right]$$

a new solution of the matrix equation A x = b.

Also from (4) for i = 1, 2, ... ,k, we have

$$\Rightarrow \qquad v \geq \frac{\lambda_i}{x_i} \text{ or } x_i \geq \frac{\lambda_i}{v}$$

$$\Rightarrow \qquad x_i - \frac{\lambda_i}{v} \geq 0.$$

Since all the components of x' are non-negative therefore x' is a feasible solution to the given L.P.P.

Also $v = \frac{\lambda_i}{x_i}$, for at least one i

i.e., $v - \frac{\lambda_i}{x_i} = 0$ therefore the new feasible solution x' cannot have more than k – 1 non-zero variables.

In this way we have derived a new feasible solution from the given optimal feasible solution which contains less number of non-zero variables. This solution is B.F.S. if the column vectors associated to non-zero variables in this new solution are L.I. If these associated vectors are not L.I., we shall repeat the whole reduction procedure as explained above. Continuing in this way for finite number of times we will derive a solution in which columns corresponding to positive variables are L.I. *i.e.,* we will obtain a B.F.S. of the system.

Now it remains to prove that x' is also an optimal solution.

Let Z' be the new value of the objective function corresponding to the new solution x'.

Then $$Z' = \sum_{i=1}^{k} c_i\left(x_i + \frac{\lambda_i}{v}\right) = \sum_{i=1}^{k} c_i x_i - \frac{1}{v}\sum_{i=1}^{1} c_i \lambda_i$$

$$\Rightarrow \quad Z' = Z^* - \frac{1}{v}\sum_{i=1}^{k} c_i \lambda_i \text{, from (2)} \qquad \text{...(6)}$$

Now for optimality, we must have

$$Z' = Z^*$$

i.e., x' will be optimal solution only if

$$\sum_{i=1}^{k} c_i \lambda_i = 0. \qquad \text{...(7)}$$

We shall prove this result by contradiction.

If possible let us assume that

$$\sum_{i=1}^{k} c_i \lambda_i \neq 0.$$

Then either $\sum_{i=1}^{k} c_i \lambda_i < 0$ ...(A)

$$\Rightarrow \quad \sum_{i=1}^{k} c_i \lambda_i > 0. \qquad \text{...(B)}$$

In either of these two cases, we can find a real number r such that $r\sum_{i=1}^{k} c_i \lambda_i > 0$ [in case (A) r will be negative and in case (B) r will be positive]

$$\Rightarrow \quad \sum_{i=1}^{k} c_i (r\lambda_i) > 0$$

$$\Rightarrow \quad \sum_{i=1}^{k} c_i (r\lambda_i) - \sum_{i=1}^{k} c_i x_i > \sum_{i=1}^{k} c_i x_i$$

$$\Rightarrow \quad \sum_{i=1}^{k} c_i(x_i + r\lambda_i) > Z^*, \text{ from (2).} \qquad ...(8)$$

Multiplying equation (3) by r and adding to (1), we get

$$\sum_{i=1}^{k} x_i\alpha_i + \sum_{i=1}^{k} r\lambda_i\alpha_i = b$$

$$\Rightarrow \quad \sum_{i=1}^{k} (x_i + r\lambda_i)\alpha_i = b.$$

Obviously $[x_1 + r\lambda_1, x_2 + r\lambda_2, \dots, x_k + r\lambda_k, 0, 0, \dots, 0]$

is also a solution of the matrix equation $A\,x = b$ for all values of r.

Furthermore, we can choose r in infinitely many ways for which the above solution also satisfies the non-negative restrictions.

Let us choose r such that

$$x_i + r\lambda_i \geq 0, \; i = 1, \dots, k$$

or $\quad r\lambda_i \geq -x_i$

$$\therefore \quad r \geq -\frac{x_i}{\lambda_i} \text{ if } \lambda_i > 0,$$

$$r \leq -\frac{x_i}{\lambda_i} \text{ if } \lambda_i < 0$$

and r is unrestricted if $\lambda_i = 0$.

Thus $x_i + r\lambda_i \geq 0$ for $i = 1, 2, \dots, k$ if we select r such that

$$\max_{\substack{i \\ (\lambda_i > 0)}} \left(-\frac{x_i}{\lambda_i}\right) \leq r \leq \min_{\substack{i \\ (\lambda_i < 0)}} \left(-\frac{x_i}{\lambda_i}\right) \qquad ...(9)$$

which means that interval given by (9) is non-empty.

Thus, r lies in the non-empty interval given in (9).

Consequently, an infinite number of solution given by

$$[x_1 + r\lambda_1, x_2 + r\lambda_2, \dots, x_k + x_k + r\lambda_k, 0, 0, \dots, 0]$$

satisfy the non-negative restrictions as well.

Now, returning to result (8) we king that $\sum_{i=1}^{k} c_i(c_i + r\lambda_i)$ yields the value of the objective function which is strictly greater than the greatest value (or optimal value) Z^* of objective function, which is impossible.

$\therefore$ we must have $\sum_{i=1}^{k} c_i \lambda_i = 0$

or $Z' = Z^*$.

Hence $x' = \left[x_i - \frac{\lambda_1}{v}, x_2 - \frac{\lambda_2}{v}, \ldots, x_k - \frac{\lambda_k}{v}, 0, 0, \ldots, 0\right]$

is also an optimal solution.

This proves the theorem.

CONDITIONS FOR THE EXISTENCE OF UNBOUNDED SOLUTIONS

Now, we have proved that if for any column aj in A but not in B the condition $c_j - Z_j > 0$ holds and if at least one $y_{ij} > 0$, i = 1, 2, ... ,m then it is possible to find an improved B.F.S.

Now the question arises : what will be the implication if for at least one a_j all $y_{ij} \leq 0$, i = 1, 2, ..., m.

To get the answer of this question we shall prove the following theorem.

Theorem:

If for any basic feasible solution $x_B = B^{-1} b$ to $A x = b$ there is some column α_j in A but not in B for which $c_j - Z_j > 0$ and $y_{ij} \leq 0$, i = 1, 2, ..., m then if the objective function is to be maximized, the problem has an unbounded solution.

Proof:

Consider the linear programming problem

Max. $Z = c\,x$ s.t. $A\,x = b$, $x \geq 0$,

where $A = (\alpha_1, \alpha_2, \ldots \alpha_N)$. $N = m + n$,

basis matrix $B = (\beta_1, \beta_2, \ldots, \beta_m)$.

Let $x_B = (x_{B1}, x_{B2}, \ldots, x_{Bm})$ be a B.F.S. of the given L.P.P. Then we have

$$Bx_B = b \text{ or } \sum_{i=1}^{m} x_{Bi}\beta_i = b \qquad \ldots(1)$$

and the corresponding value of the objective function is

$$Z = c_B\ x_B = \sum_{i=1}^{m} c_{Bi} x_{Bi} \qquad ...(2)$$

Adding and subtracting $\lambda\alpha_j$ in (1), we get

$$\sum_{i=1}^{m} x_{Bi}\beta_i - \lambda\alpha_j + \lambda\alpha_j = b, \qquad ...(3)$$

where l is some scalar.

Let $\alpha_j \in A$ s.t. $\alpha_j \notin B$.

Since the vectors $\beta_1, \beta_2, ..., \beta_m$ are in the basis of A therefore we can express α_j as the linear combination of b's

$$\therefore \quad \alpha j = \sum_{i=1}^{m} y_{ij}\beta_i$$

$$\Rightarrow -la_j = -1 \sum_{i=1}^{m} y_{ij}\beta_i .$$

Substituting the above value of $-\lambda\alpha_j$ in (3), we get

$$\Rightarrow \quad \sum_{i=1}^{m} x_{Bi}\beta_i - \lambda \sum_{i=1}^{m} y_{ij}\beta_i + \lambda\alpha_j = b$$

$$\Rightarrow \quad \sum_{i=1}^{m} (x_{Bi} - \lambda y_{ij})\beta_i + \lambda\alpha_j = b$$

Thus we obtain a new solution

$$x'_B = (x_{B1}', x_{B2}', ... ,x_{Bm}', \lambda) \qquad ...(4)$$

where $x_{Bi}' = x_{Bi} - \lambda y_{ij}$, $i = 1, 2, ..., m$.

When $\lambda > 0$, $x_{Bi} - \lambda y_{ij} \geq 0$ since $y_{ij} \leq 0$, $i = 1, 2, ... ,m$ therefore (4) gives the feasible solution in which the number of positive variables is less than or equal to (m + 1). It may be less than (m + 1) because some $x_{Bi} - \lambda y_{ij}$ may be zero for some i. In case the number of positive variables in this solution is equal to (m + 1) then this solution will be non-basic feasible solution.

If Z' is the new value of the objective function corresponding to this new solution, then we have

$$\Rightarrow \quad Z' = \sum_{i=1}^{m} c_{Bi}(x_{Bi} - \lambda y_{ij}) + c_j\lambda$$

$$\Rightarrow \quad Z' = \sum_{i=1}^{m} c_{Bi}x_{Bi} + \lambda(c_j - c_{Bi}y_{ij}) = Z + \lambda\ (c_j - Z_j).$$

Since $c_j - Z_j > 0$ (given) therefore Z' can be made as large as we please by giving sufficiently large values to l. We know that a L.P.P. has an unbounded solution if the value of its objective function can be increased or decreased arbitrarily.

Hence the given L.P.P. has an unbounded solution.

ARTIFICIAL VARIABLES TECHNIQUE

(a) **Big-M Method :** Sometimes in linear programming problems constraints may also have ≥ and = signs after ensuring that all bi ≥ 0. In such problems we first introduce *surplus variables* to convert inequalities into equations. In such cases basis matrix is not obtained as an identity matrix in the starting simplex table. To overcome this difficulty we introduce new variables, called, the *artificial variables* to each of such constraints. These variables are fictitious and cannot have any physical meaning.

The artificial variables are introduced for the limited purpose of obtaining an *initial solution.* It is not relevant whether the objective function is of the maximization or minimization type. Since artificial variables do not represent any quantity relating to the decision problem they must be driven out of the system and must not be present in the final solution. Keeping this in mind a method known as *big-M method or Cranes M-method* was developed. It we suggestedby A. Cranes that a very high penalty can be paid for introducing the artificial variables in the constraints of a given problem by assigning an extremely high negative cost (penalty) to them.

∴ the objective function can be written as

$$Z = c\,x + 0.x_{slack} + 0.x_{surplus} - M.x_{art}.$$

It is significant to note that the initial solution obtained using the artificial variables is not a feasible solution to the given problem. It only gives the starting point and the artificial variables are driven out in he normal course of applying simplex method to the problem.

Note: A solution to the problem which does not contain an artificial variables in the basic, represents a feasible solution to the problem.

BASIC FEASIBLE SOLUTION TO BECOME OPTIMAL

Theorem:

Let us $x_B = B^{-1} b$ be the basic (degenerate or non-degenerate) feasible solution of the L.P.P. Z = c x s.t. A x = b, x ≥ 0. Let $Z^* = c_B\, x_B$ be the value of the objective function at any iteration of simplex method. If $c_j - Z_j \le 0$ for every column α_j in A but not in B then Z* is the optimum value of the objective function Z and x_B is an optimal basic feasible solution.

Proof:

Suppose $x_B = (x_{B1}, x_{B2}, \dots, x_{Bm})$ is the B.F.S. of the given L.P.P. We have,

$$A = (\alpha_1, \alpha_2, \dots, \alpha_N),\ N = m + n$$

and the basis matrix $B = (\beta_1, \beta_2, \dots, \beta_m)$.

$$\therefore \qquad B_{mB} = b \text{ or } x_B = B^{-1} b. \qquad \text{...(1)}$$

The value of the objective function for the B.F.S. xB is

$$Z^* = c_B\, x_B = \sum_{i=1}^{m} c_{Bi} x_{Bi}\,. \qquad \text{...(2)}$$

Also, we are given that $c_j - Z_j \le 0$ for every column α_j in A but not in B.

Let 'x' = $[x_1', x_2', \dots, x_N']$ be any feasible solution and Z' be the value of the objective function for this solution.

Then we have $A\,x' = b$ ...(3)

$$\text{and} \qquad Z' = c\,x' = \sum_{j=1}^{N} c_j x_j' \qquad \text{...(4)}$$

Now Z* will be the optimum value of the objective function if we have $Z^* \ge Z'$. To prove $Z^* \ge Z'$ we proceed in the following manner.

We have $x_B = B^{-1} b = B^{-1} (A\,x')$, from (3)

$$= (B^{-1} A)\, x'$$

$$= Y\,x', \text{ where } Y = [Y_1, Y_2, \dots, Y_N]$$

$$= [Y_1, Y_2, \dots, Y_N]\,[x_1', x_2', \dots, x_N']$$

$$= \begin{bmatrix} y_{11} & y_{12} & \cdots & y_{1j} & \cdots & y_{1N} \\ y_{21} & y_{22} & \cdots & y_{2j} & \cdots & y_{2N} \\ \cdots & \cdots & \cdots & \cdots & \cdots & \cdots \\ \cdots & \cdots & \cdots & \cdots & \cdots & \cdots \\ y_{m1} & y_{m2} & \cdots & y_{mj} & \cdots & y_{mN} \end{bmatrix} \begin{Bmatrix} x_1' \\ x_2' \\ \vdots \\ x_j' \\ \vdots \\ x_N' \end{Bmatrix}$$

$\Rightarrow \qquad [x_{B1}, x_{B2}, ..., x_{Bi}, ..., x_{Bm}]$

$$= \left[\sum_{j=1}^{N} y_{1j}x_j', \sum_{j=1}^{N} y_{2j}x_j', ..., \sum_{j=1}^{N} y_{ij}x_j', ..., \sum_{j=1}^{N} y_{mj}x_j'\right]$$

Equating ith component on both sides, we get,

$$x_{Bi} = \sum_{j=1}^{N} y_{ij}x_j'. \qquad ...(5)$$

We have $c_j - Z_j \le 0$ for all j for which α_j is not in B.

If this inequality also holds for all those j for which α_j is in B then $c_j - Z_j \le 0$ for all j for which α_j also belong to B.

Now if $\alpha_j = B_i$, then

$$\alpha_j = \beta_i = 0.\beta_1 + 0.\beta_2 + ... + 0.\beta_{i-1} + 1.\beta_i + ... + 0.\beta_m$$

which shows that $Y_j = e_i$, a vector whose ith component is unity.

Again $\quad \alpha_j = \beta_i$ gives $c_j = c_{Bi}$

$\therefore \qquad c_j - Z_j = c_j - c_B Y_j = c_j - c_B e_i = c_j - c_{Bi} = 0$

Thus $c_j - Z_j = 0$ for all those j for which $\alpha_j \in$ B.

$\therefore \qquad c_j - Z_j \le 0$ for all α_j in A

$\Rightarrow \qquad c_j \le Z_j$

$\Rightarrow \qquad c_j x_j' \le Z_j x_j', \because x_j' \ge 0$

$$\Rightarrow \qquad \sum_{i=1}^{N} c_j x_j' \le \sum_{j=1}^{N} Z_j x_j'$$

$$\Rightarrow \qquad Z' \le \sum_{j=1}^{N} x_j'(c_B Y_j), \text{ from (4)}$$

$$= \sum_{j=1}^{N} x_j' \left(\sum_{i=1}^{m} c_{Bi} y_{ij}\right)$$

$$= \sum_{i=1}^{m} c_{Bi} \left(\sum_{j=1}^{N} x_j' y_{ij}\right)$$

$$= \sum_{i=1}^{m} c_{Bi} x_{Bi}, \text{ from (5)}$$

$$\therefore Z' \le Z^*.$$

Hence Z* is the optimum value of the objective function.

THE ALTERNATIVE OPTIMAL SOLUTIONS

The given linear programming problem is said to possess an alternative optimal solution if the set of variables giving the optimal value of the objective function is not unique.

Theorem:

(Conditions for Alternative Optimum Solution)

Suppose there exists an optimum B.F.S. to a L.P.P. and

(i) if for some a_j in A but not in B, $c_j - Z_j = 0$ and y_{ij} £ 0 for all $i = 1, 2, ..., m$ then a non-basic alternative solution will exist.

(ii) if for some a_j in A but not in B, $c_j - Z_j = 0$ and $y_{ij} > 0$ for all least one i then an alternative basic optimum solution will exist.

Proof:

(i) We have discussed in 6 that if we introduce the column vector a_j in B where a_j is in A but not in B and $y_{ij} \leq 0$ for all $i = 1, 2, ... ,m$ then a non-basic feasible solution x_B' with (m + 1) number of positive variables is given by

$$\Rightarrow \qquad x_B' = [x_{B1}', x_{B2}', ... ,x_{Bm}', l]$$

where $$x_{Bi}' = x_{Bi} - \lambda y_{ij}, \; i = 1, 2, ... ,m, \; l > 0$$

and the value of the objective function for this new F.S. is given by

$$Z' = Z + l\,(c_j - Z_j).$$

If $$c_j - Z_j = 0, \text{ then } Z' = Z$$

i.e., the value of the objective function for this new non-basic F.S. is also equal to Z (optimal value). Hence this new non-basic F.S. is an alternative optimal solution of the given L.P.P.

(ii) We have shown in theorem of 5 that if $y_{ij} > 0$ for at least one $i = 1, 2, ..., m$ then by palpation one column br in B by the column aj which is in A but not in B, we obtain a new B.F.S. x_B' given by

$$x_B' = [x_{B1}', x_{B2}', ... ,x_{Bm}']$$

where $$x_{Bi}' = x_{Bi} - \frac{y_{ij}}{y_{rj}} x_{Br}, \; i = 1, 2, ... ,m, \; i \geq r$$

$$x_{Br}' = \frac{x_{Br}}{y_{rj}}, \; i = r$$

and $\frac{x_{Br}}{y_{rj}} = \underset{i}{\text{Mini}}\left(\frac{x_{Bi}}{y_{ij}}, y_{ij} > 0\right)$.

The value of objective function for this new B.F.S. is given by

$$Z' = Z + \frac{x_{Br}}{y_{rj}}(c_j - Z_j)$$

i.e., the value of the objective function for this new B.F.S. is also equal to Z' (optimal value).

Hence this new B.F.S. is an alternative optimal F.F.S.

TO DETERMINE STARTING B.F.S.

In this article we are going to discuss a method for finding a most convenient initial B.F.S. to a linear programming problem .

In the constraints of a general linear programming problem three may be any of the three signs $\leq = \geq$. Let us assume that the requirement vector $b \geq 0$ (if any of the b_i's is negative, multiply the corresponding constraint by – 1.)

Case I. To find initial B.F.S. when at the original constraints have $\leq$ sign. To convert all the constraints into equations we insert slack variables only .the equations obtained are as follows.

$$a_{11} x_1 + a_{12} x_2 + \ldots + a_{1n} x_n + 1.x_{n+1} = b_1$$

$$a_{21} x_1 + a_{22} x_2 + \ldots + a_{2n} x_n + 1.x_{n+2} = b_2$$

$$\ldots \quad \ldots \quad \ldots \quad \ldots \quad \ldots \quad \ldots$$

$$\ldots \quad \ldots \quad \ldots \quad \ldots \quad \ldots \quad \ldots$$

$$a_{m1} x_1 + a_{m2} x_2 + \ldots + a_{mn} x_n + \ldots + 1.x_{n+m} = b_m$$

Here x_{n+1}, x– = 2, ... ,x_{n+m} are slack variables.

In matrix from these equations can be written as

$$\begin{bmatrix} a_{11} & a_{12} & \cdots & a_{1n} & 1 & 0 & \cdots & 0 \\ a_{21} & a_{22} & \cdots & a_{2n} & 0 & 1 & \cdots & 0 \\ \cdots & \cdots & \cdots & \cdots & \cdots & & \cdots & \\ \cdots & \cdots & \cdots & \cdots & \cdots & & \cdots & \\ a_{m1} & a_{m2} & \cdots & a_{mn} & 0 & 0 & \cdots & 1 \end{bmatrix} \begin{bmatrix} x_1 \\ x_2 \\ \vdots \\ x_n \\ x_{n+1} \\ \vdots \\ x_{n+m} \end{bmatrix} = \begin{bmatrix} b_1 \\ b_2 \\ \vdots \\ b_m \end{bmatrix}$$

Taking the initial basis matrix

$$B = I_m, \text{ (unit matrix of order } m \times m).$$

$\therefore$ the initial basic solution is given by $x_B = B^{-1} b = I_m b = b \geq 0$

Thus the initial B.F.S. is

$$x_{n+1} = x_{B1} = b_1, x_{n+2} = x_{B2} = b_2, \ldots, x_{n+m} = x_{Bm} = b_m,$$

which can be obtained by writing all the non-basic variables (*i.e.*, given variables) $x_1, x_2, \ldots x_n$ equal to zero and solving the equations for the remaining variables (*i.e.*, slack variables) $x_{n+1}, x_{n+2}, \ldots, x_{n+m}$.

Case II. To find initial B.F.S. when all the original constraints have $\geq$ sign. First we convert all the constraints into equations by inserting surplus variables. The equations obtained are as follows:

$$\begin{array}{l} a_{11} x_1 + a_{12} x_2 + \ldots + a_{1n} x_n - x_{n+1} = b_1 \\ a_{21} x_1 + a_{22} x_2 + \ldots + a_{2n} x_n - x_{n+2} = b_2 \\ \ldots \quad \ldots \quad \ldots \quad \ldots \quad \ldots \quad \ldots \\ \ldots \quad \ldots \quad \ldots \quad \ldots \quad \ldots \quad \ldots \\ a_{m1} x_1 + a_{m2} x_2 + \ldots + a_{mn} x_n - x_{n+m} = b_m \end{array}$$

Here $x_{n+1}, x_{n+2}, \ldots, x_{n+m}$ are surplus variables.

In matrix form, these equations can be writing as

$$\begin{bmatrix} a_{11} & a_{12} & \ldots & a_{1n} & -1 & 0 & \ldots & 0 \\ a_{21} & a_{22} & \ldots & a_{2n} & 0 & -1 & \ldots & 0 \\ \ldots & \ldots & & & \ldots & \ldots & & \ldots \\ \ldots & \ldots & & & \ldots & \ldots & & \ldots \\ a_{m1} & a_{m2} & 0 & 0 & \ldots & & \ldots & -1 \end{bmatrix} \begin{bmatrix} x_1 \\ x_2 \\ \vdots \\ x_n \\ x_{n+1} \\ \vdots \\ x_{n+m} \end{bmatrix} = \begin{bmatrix} b_1 \\ b_2 \\ \vdots \\ b_m \end{bmatrix}$$

If we take the initial basis matrix $B = -I_m$,

we have $x_B = B^{-1} b = = -I_m b = -b \leq 0$, which is not a B.F.S.

To avoid this difficulty we add one more variable to each constraint. These variables are called *"Artificial Variables"*.

Adding surplus and artificial variables, the constraints of the given L.P.P. change to the following equations.

$$\begin{array}{l} a_{11} x_1 + a_{12} x_2 + \ldots + a_{1n} x_n - x_{n+1} + x_{n+m+1} = b_1 \\ a_{21} x_1 + a_{22} x_2 + \ldots + a_{2n} x_n - x_{n+2} + x_{n+m+2} = b_2 \end{array}$$

$$\begin{array}{ccccccc} \dots & \dots & \dots & \dots & \dots & \dots & \dots \\ \dots & \dots & \dots & \dots & \dots & \dots & \dots \end{array}$$

$$a_{m1} x_1 + a_{m2} x_2 + \dots + a_{mn} x_n - x_{n+m} + m_{n+m+m} = b_m$$

Here $x_{n+m+1}, x_{n+m+2}, \dots, x_{n+m+m}$ are the artificial variables.

In matrix, these equations can be written as

$$\begin{bmatrix} A_{11} & a_{12} & \dots & a_{1n} & -1 & 0 & \dots & 0 & 1 & 0 & \dots & 0 \\ a_{21} & a_{22} & \dots & a_{2n} & 0 & -1 & \dots & 0 & 0 & 1 & \dots & 0 \\ \dots & \dots & \dots & \dots & \dots & \dots & \dots & \dots & \dots & \dots & \dots & \dots \\ \dots & \dots & \dots & \dots & \dots & \dots & \dots & \dots & \dots & \dots & \dots & \dots \\ a_{m1} & a_{m2} & \dots & a_{mn} & 0 & 0 & \dots & -1 & 0 & 0 & \dots & 1 \end{bmatrix}$$

$$\begin{bmatrix} x_1 \\ x_2 \\ \vdots \\ x_n \\ x_{n+1} \\ \vdots \\ x_{n+m} \\ x_{n+m+1} \\ \vdots \\ x_{n+m+m} \end{bmatrix} = \begin{bmatrix} b_1 \\ b_2 \\ \vdots \\ b_m \end{bmatrix}$$

Now taking the basis matrix $B = I_m$.

$$\therefore \qquad x_B = B^{-1} b = I_m b = b \geq 0,$$

which is a basic feasible solution.

$\therefore$ the B.F.S. is $x_{n+m+1} = x_{B1} = b_1, x_{n+m+2} = x_{B2} = b_2, \dots, x_{n+m+m} = x_{Bm} = b_m$, which can be obtained by writing all the non basic variables (*i.e.*, given variables and surplus variables) $x_1, x_2, \dots, x_n; x_{n+1}, x_{n+2}, \dots, x_{n+m}$ equal to zero and solving the equations for the remaining basic variables (*i.e.*, artificial variables) $x_{n+m+1}, x_{n+m+2}, \dots, x_{n+m+m}$.

Case III. To find initial B.F.S. when the constraints have '≤', '≥' and '=' signs. In this case we convert the constraint into equations by inserting slack, surplus and artificial variables. Here the basis matrix $B = I_m$ is obtained by introducing unit column vectors corresponding to the slack and artificial variables. To obtain the initial B.F.S. we put the non-basic variables equal to zero and solve for remaining basic variables. Here also the initial B.F.S. $x_B = b \geq 0$.

COMPUTATIONAL PROCEDURE OF THE SIMPLEX METHOD

Step 1. Find the initial B.F.S. follow the method discussed in 10 to find the initial B.F.S. If artificial variables are introduced in the L.P.P. then follow the two-method for Big M-method for solving such prob lems.

Step 2. Construct the initial simplex table (starting simplex table).

It should be remembered that the values of non-basic variables are always zero at each iteration

Step 3. Test the initial B.F.S. for optimality.

Compute the net evaluation Δ_j for each variables x_j by using the formula $\Delta_j = c_j - c_B Y_j$.

Step 4. If the given problem is of minimization, convert it into the maximization problem.

For this, multiply both sides of the objective function by –1 and put $-Z = Z'$.

If v is the maximum value of Z' then –v will be the minimum value of Z.

Step 5. Make all the b_i's non-negative.

The R.H.S. of each of the constraints should be non-negative. If the L.P.P. has a constraint for which a negative b_i is given, it should be converted into positive value by multiplying both sides of the constraint by –1.

Step 6. Convert inequalities of constraints into equations.

For this introduce slack or surplus variables. The coefficients of slack or surplus variables in the objective function are zero. Introduce artificial variables, if necessary.

OPTIMAL TEST

(a) If $\Delta_j \leq 0$ for all j, the solution under consideration is optimal.

Alternative optimal solution will exist if any non-basic Δ_j is also zero.

(b) If $\Delta_j > 0$ for any j *i.e.*, if at least one Δ_j is positive the solution under test is not optimal, we must proceed to improve the solution (step 7).

(c) If corresponding to maximum positive D_j, all elements of the column Y_j are negative or zero, the solution under test will be unbounded.

(d) If optimality condition is satisfied but the value of at least one artificial variables present in the basis is non-zero, the problem will have no feasible solution.

Step 1. *Select the entering (incoming) vector and departing (outgoing) vector.*

To improve the initial B.F.S. we select the vector entering the basis matrix and the vector to be removed from the basis matrix.

Incoming Vector. The incoming vector α_k is always selected corresponding to the largest positive value of Δ_j.

If maximum value of Δ_j occurs for more than one α_j then we can select any of these vectors as incoming vector.

Outgoing Vector. The departing vector β_r is selected corresponding to that value of r for which

$$\frac{x_{Br}}{y_{rk}} = \min_i \left\{ \frac{x_{Bi}}{y_{ik}}, y_{ik} > 0 \right\},$$

is α_r is the incoming vector.

Step 2. If α_k is the entering vector and β_r is the outgoing vector then the element yrk which lies at the intersection of minimum ration arrow (←) and incoming vector arrow ↑ is called the pivot element (key element).

We put this element in □.

In order to bring b_r in place of incoming vector Y_k, unity must occupy the place □. In other words key element y_{rk} should be 1. If it is not so then divide all the elements of this row by the key element y_{rk}. Then subtract suitable multiples of the row containing the key element from all other rows and obtain zero at all other positions of this column Y_k. Now bring br in place of α_r and construct the revised simplex table.

Thus, we get improve basic feasible solution.

Step 3. Test the improved B.F.S. for optimality.

If it is not optimum, repeat steps 7 and 8 until we obtain an optimums solution.

Note: If the above mentioned minimum value is not unique then more than one variable will vanish in the next solution. As a result the next solution will be a degenerate B.F.S. for which the outgoing vector is selected in a different way.

COMPUTE THE INVERSE OF A MATRIX WHOSE ONE COLUMN IS DIFFERENT FORM OF A MATRIX WHOSE INVERSE IS KNOWN

$B = (\beta_1, \beta_2, \ldots, \beta_n)$ is the given non-singular matrix whose inverse is known.

Consider a matrix $B_\alpha = (\beta_1, \beta_2, \ldots, \beta_{r-1}, \alpha, \beta_{r+1}, \ldots, \beta_n)$.

We have obtained the matrix Ba by replacing the rth column of matrix B by the column vector a.

If we have $\alpha = \sum_{i=1}^{n} y_i \beta_i$,

Ba will be non-singular iff $y_r \neq 0$.

Using components of the vector $Y = B^{-1}\alpha$, we introduce a new matrix E where E is a square matrix of order n which differs from identity matrix In the rth column only. To obtain the rth column of matrix E we proceed in the following manner:

We have $\alpha = y_1 \beta_1 + y_2 \beta_2 + \ldots + y_r \beta_r + \ldots + y_m \beta_m$.

Transposing yr br to the left and a to the right, we get

$$-y_r \beta_r = y_1 \beta_1 + y_2 \beta_2 + \ldots + (-\alpha) + \ldots + y_n \beta_n$$

$$\text{or} \quad \beta_r = -\frac{y_1}{y_r}\beta_1 - \frac{y_2}{y_r}\beta_2 - \ldots + \frac{1}{y_r}\alpha - \ldots - \frac{y_n}{y_r}\beta_n$$

$$= \frac{y_1}{y_r}\beta_1 - \frac{y_2}{y_r}\beta_2 - \ldots + \frac{1}{y_r}\alpha - \ldots - \frac{y_n}{y_r}\beta_n \; (\beta_1, \beta_2, \ldots, \alpha, \ldots \beta_n)$$

$$= A_r B_\alpha,$$

where $A_r = \left(-\frac{y_1}{y_r}, -\frac{y_2}{y_r}, \ldots, \frac{1}{y_r}, -\frac{y_n}{y_r}\right)$ gives the rth column of E.

$$\begin{aligned} \text{Now } B_\alpha E &= B_\alpha (e_1, e_2, \ldots, e_{r-1}, A_r, e_{r+1}, \ldots, e_n) \\ &= (B_\alpha e_1, B_\alpha e_2, \ldots, B_\alpha e_{r-1}, B_\alpha A_r, \ldots, B_\alpha e_n) \\ &= (\beta_1, \beta_2, \ldots, \beta_{r-1}, \beta_r, \beta_{r+1}, \ldots, \beta_n), \end{aligned}$$

since $B_\alpha e_1 = (\beta_1, \beta_2, \ldots, \alpha, \ldots, \beta_n)(1, 0, 0, \ldots, 0)$

$$= \beta_1 + 0 + 0 + \ldots + 0 = \beta_1 \text{ etc.} = B.$$

Since $\quad B = B\alpha\, E$

$\therefore \quad BB^{-1} = B_\alpha\, EB^{-1} \quad \Rightarrow \quad I = B_\alpha\, EB^{-1}$

$\Rightarrow \quad B_\alpha^{-1} I = (B_\alpha^{-1} B_\alpha)\, EB^{-1} \quad \Rightarrow \quad B_\alpha^{-1} = IEB^{-1} = EB^{-1}.$

SOLVED EXAMPLES

Example 1:

Maximize $Z = 2x_1 + 4x_2$

subject to $2x_1 + x_2 \leq 18$

$3x_1 + 2x_2 \geq 30$

$x_1 + 2x_2 = 26;\ x_1, x_2 \geq 0.$

Solution:

The problem is of maximization and all b_i's are positive.

Introducing the necessary slack, surplus and artificial variables, the problem reduces to the following form:

$$\text{Max. } Z = 2x_1 + 4x_2 + 0x_3 - 0x_4 - MA_1 + MA_2$$

subject to $2x_1 + x_2 + x_3 = 18$

$3x_1 + 2x_2 + 0x_3 - x_4 + A_1 = 30,\ x_1 + 2x_2 + 0x_3 + 0x_4 + A_2 = 26$

The starting B.F.S. is

$x_1 = 0,\ x_2 = 0,\ x_3 = 18,\ x_4 = 0,\ A_1 = 30,\ A_2 = 26.$

The solution to the problem using simplex algorithm is given below:

First Simplex Table

		c_j	2	4	0	0	–M	–M	Min.
ratio									
B	c_B	x_B	Y_1	Y_2	Y_3	Y_4	A_1	A_2	x_B/Y_2
Y_3	0	18	2	1	1	0	0	0	18
A_1	–M	30	3	2	0	–1	1	0	15
A_2	–M	26	1	(2)	0	0	0	1	13 ←
$Z=c_B\, x_B= -56M$		Δ_j	2+4M	4+4M ↑	0	–M	0	0 ↓	

Computing Δ_j using the formula $\Delta_j = c_j - c_B Y_j$, we get

$$\Delta_1 = c_1 - c_B Y_1 = 2 - (0, -M, -M)(2, 3, 1) = 2 + 4M$$

$$\Delta_2 = c_2 - c_B Y_2 = 4 - (0, -M, -M)(1, 2, 2) = 4 + 4M$$

$$\Delta_4 = c_4 - c_B Y_4 = 0 - (0, -M, -M)(0, -1, 0) = -M.$$

Y_2 is the entering vector. Using min. ratio rule we find that 2 is the pivot element which indicates that A_2 is the outgoing vector.

Second Simplex Table

		c_j	2	4	0	0	–M	–M	Min. ratio
B	c_B	x_B	Y_1	Y_2	Y_3	Y_4	A_1	A_2	x_B/Y_1
Y_3	0	5	3/2	0	1	0	0	–1/2	10/3
A_1	–M	4	(2)	0	0	–1	0	–1	2 ←
Y_2	4	13	1/2	1	0	0	0	1/2	26
Z = 52 – 4M		Δ_j	2M ↑	0	0	–M	0 ↓	–2 – 2M	

Y_1 is the incoming vector and A_1 is the outgoing vector.

Third Simplex Table

		c_j	2	4	0	0	–M	–M
B	c_B	x_B	Y_1	Y_2	Y_3	Y_4	A_1	A_2
Y_3	0	2	0	0	1	3/4	–3/4	1/4
Y_1	2	2	1	0	0	–1/2	1/2	–1/2
Y_2	4	12	0	1	0	1/4	–1/4	3/4
Z=52	Δ_j	0	0	0	0	–M	–M	

Since all Δ_j are negative or zero therefore the solution obtained is optimal.

∴ the optimal solution is

$x_1 = 2$, $x_2 = 12$, $x_3 = 2$ and other variables = 0.

The objective function Z = 52.

Example 2:

Minimize $Z = x_1 + x_2 + x_3$

subject to

$x_1 - 3x_2 + 4x_3 = 5$

$x_1 - 2x_2 \leq 3$

$2x_2 - x_3 \geq 4,$

$x_1, x_2 \geq 0$ *and* x_3 *is unrestricted.*

Solution:

First we shall convert the minimization problem into maximization by substituting Z' = – Z.

The objective function changes to

$$\text{Max. } Z' = -x_1 - x_1 - x_3.$$

Since x_3 is unrestricted therefore replacing it by

$$x_3' - x_3'' \text{ where } x_3', x_3'' \geq 0.$$

Also introducing slack variables x_4, surplus variables x_5 and artificial variables A_1 and A_2, the given problem becomes

$$\text{Max } Z' = -x_1 - x_2 - x_3' + x_3'' + 0x_4 + 0x_4 + 0x_5 - MA_1 - MA_2$$

s.t. $x_1 - 3x_2 + 4x_3' - 4x_3'' + A_1 = 5$

$x_1 - 2x_2 + x_4 = 3$

$0x_1 + 2x_2 - x_3' + x_3'' - x_5 + A_2 = 4,$

$x_1, x_2, x_3', x_3'', x_4, x_5, A_1, A_2 \geq 0.$

Now solve the problem using simplex algorithm.

The optimal solution is

$\Rightarrow \quad x_1 = 0, x_2 = 21/5,$

$\Rightarrow \quad x_3 = 22/5$

and Min. Z = 43/5.

Example 3:

Maximize $Z = 40x_1 + 35x_2$

subject to $2x_1 + 3x_2 \leq 60$

$4x_1 + 2x_2 \leq 96,$

$x_1, x_2 \geq 0.$

For the clear understanding of the simplex method the solution to this problem is illustrated below in a step-wise manner.

Solution:

Step 1. The given problem is of maximization and all the b_i's are already positive.

Step 2. Converting the inequalities of constraints into equations by introducing slack variables x_3 and x_4, we get

$$2x_1 + 3x_2 + x_3 = 60 \text{ and } 4x_1 + 3x_2 + x_4 = 96.$$

The coefficients of slack variables are zero in the objective function. Therefore the given problem becomes

$$\text{Maximize } Z = 40x_1 + 35x_2 + 0x_3 + 0x_4$$

$$\text{subject to } 2x_1 + 3x_2 + x_3 + 0x_4 = 60$$

$$4x_1 + 3x_2 + 0x_3 + x_4 = 96$$

$$x_1, x_2, x_3, x_4 \geq 0.$$

This is the standardized form of the given problem.

Step 3. The simplex method begins with an initial basic feasible solution in which the values of basic variables are zero.

$\therefore$ $x_1 = 0$, $x_2 =$ which gives $x_3 = 60$, $x_4 = 96$.

Step 4. Constructing the initial simplex table.

First Simplex Table

		c_j	40	35	0	0	Min. ratio
B	c_B	x_B	Y_1	Y_2	Y_3	Y_4	x_B/Y_1
Y_3	0	60	2	3	1	0	60/2 = 30
Y_4	0	96	(4)	3	0	1	96/4 = 24
	←						
$Z = c_B\ x_B = 0$		Δ_j	40 ↑	35	0	0 ↓	

Step 5. Computing Δ_j for all non-basic variables x_j, j = 1, 2, using the formula $\Delta_j = c_j - c_B Y_j$

$$\Delta_1 = c_1 - c_B Y_1 = 40 - (0, 0)(2, 4) = 40$$

$$\Delta_2 = c_2 - c_B Y_2 = 35 - (0, 0)(3, 3) = 35.$$

Since all Δ_j's are not less than or equal to zero, therefore the solution is not optimal. So we proceed to next step to improve the solution.

Step 6. Incoming and outgoing vectors.

Since $\Delta_1 = 40$ is the maximum value of Δ_j, j = 1, 2, therefore $\alpha_1 = Y_1$ is the incoming vector.

$$\text{Now } \frac{x_B}{Y_1} = \left(\frac{x_{B1}}{y_{11}}, \frac{x_{B2}}{y_{21}}\right) = \left(\frac{60}{2}, \frac{96}{4}\right) = (30, 24).$$

$$\frac{x_{Br}}{y_{r1}} = \min_i \left[\frac{x_{Bi}}{y_{i1}}, y_{i1} > 0\right] = \min_i \left(\frac{x_{B1}}{y_{11}}, \frac{x_{B2}}{y_{21}}\right) = 24$$

$$= \frac{x_{B2}}{y_{21}}$$

$\therefore$ r = 2 which gives $\beta_2 = Y_4$ as the outgoing vector.

$\therefore$ $y_{21} = \alpha_{21}$ is the key element which is equal to 4.

To bring Y_1 in place of Y_4 we proceed in the following manner.

Dividing the second row containing the key element $\alpha_{21} = 4$ by 4 to make if unity.

Now, we shall subtract appropriate multiples of this new row form the other remaining row so as to obtain zeros in the remaining positions of Y_1.

Subtitling 2 times of the second row from the first row and subtract 40 times of the second row from the row containing Δ_j's to obtain new values of Δ_j's.

Constructing the second simplex table in which β_2 is replaced by α_1.

Second Simplex Table

		c_j	10	35	0	0	Min ratio
B	c_B	x_B	Y_1	Y_2	Y_3	Y_4	x_B/Y_2
Y_3	0	12	0	(3/2)	1	−1/2	8 ←
Y_1	40	24	1	3/4	0	1/4	32
$Z = c_B\ x_B = 960$		Δ_j	0	5 ↑	0 ↓	−10	

Also computing Δ_j by using the formula $\Delta = c_j - c_B Y_1$ for x_2, x_4, we get

$$\Delta_2 = c_2 - c_B Y_2 = 35 - (0, 40)\left(\frac{3}{2}, \frac{3}{4}\right) = 5$$

$$\Delta_4 = c_4 - c_B Y_4 = 0 - (0, 40)\left(-\frac{1}{2}, \frac{1}{4}\right) = -10.$$

Δ_1 and Δ_3 are zero as they correspond to unit column vectors.

Since all Δ_j's are not less than or equal to zero therefore the solution obtained is not optimal.

$\Delta_2 = 5$ is the maximum value of Δ_j.

$\therefore \quad \alpha_2 = Y_2$ is the incoming vector.

$$\text{Now } \frac{x_{Br}}{y_{r2}} = \min_i \left[\frac{x_{Bi}}{y_{i2}}, y_{i2} > 0\right]$$

$$= \min\left[\frac{x_{B1}}{y_{12}}, \frac{x_{B2}}{y_{22}}\right] = \min\,[8, 32] = 8$$

$$= Y_3$$

$\alpha_{12} = y_{12} = \frac{3}{2}$ is the key element.

To bring Y_2 in place of Y_3 we proceed in he following manner.

Dividing the first row by 3/2 to make the key element unity.

Subtracting 3/4 times of the first row from the second row and subtracting 5 times of the list row form the two containing Δ_j's to obtain new values of Δ_j's.

Constructing the third simplex table in which Y_3 is replaced by Y_2.

Third Simplex Table

B	c_B	c_j / x_B	40 / Y_1	35 / Y_2	0 / Y_3	0 / Y_4	Min. ratio
Y_2	35	8	0	1	2/3	–1/3	
Y_1	40	18	1	0	–1/2	1/2	
$Z=c_B$	x_B=1000	Δ_j	0	0	–10/3	–25/3	

Since all the Δ_j's are less than or equal to zero therefore the above solution is optimal.

Hence the optimal solution is

$$x_1 = 18,\ x_2 = 8 \text{ and max. } Z = 1000.$$

The above solution with its different computational steps can be more conveniently represented by a single table as shown below:

		c_j	40	35	0	0	Min. ratio
B	c_B	x_B	Y_1	Y_2	Y_3	Y_4	x_B/Y_1
Y_3	0	60	2	3	1	0	60/2 = 30
Y_4	0	96	(40)	3	0	1	96/4 = 24
$Z=c_B$	$x_B=0$	Δ_j	40 ↑	35	0	0 ↓	x_B/y_2
Y_3	0	12	0	(3/2)	1	–1/2	8
Y_1	40	24	1	3/4	0	1/4	32
$Z=c_B$	$x_B=960$	Δ_j	0	5 ↑	0 ↓	–10	
Y_2	35	8	1	1	2/3	–1/3	
Y_1	40	18	1	0	–1/2	1/2	
$Z=c_B$	$x_B=1000$	Δ_j	0	0	–10/3	–25/3	

Example 4:

Determine the value of u_1, u_2 and u_3 so as to

$$\textit{Maximize } Z = u_1 \cdot u_2 \cdot u_3$$

subject to the constraint

$$u_1 + u_2 + u_3 = 10$$

and $u_1, u_2, u_3 \geq 0$.

Solution:

Let us define state variable x_j (j = 1, 2, 3) such that

$x_3 = u_1 + u_2 + u_3 = 10$, at stage 3

$x_2 = x_3 - x_3 = u_1 + u_2$, at stage 2

$x_1 = x_2 - u_2 = u_1$, at stage 1

The maximum value of Z for any feasible value of state variable is given by

$$f_3(x_3) = \underset{u_3}{\text{Max}} \{u_3 \cdot f_2(x_2)\}$$

$$f_2(x_2) = \underset{u_3}{\text{Max}} \{u_2 \cdot f_1(x_1)\}$$

$$f_1(x_1) = u_1 = x_2 - u_2$$

Thus $f_2(x_2) = \underset{u_3}{\text{Max}} \{u_2 \cdot (x_2 - u_2)\} = \underset{u_3}{\text{Max}} \{u_2 x_2 - u_2^2\}$

Differentiating $f_2(x_2)$ with respect to u_2 and equating to zero (necessary condition for maximum or minimum value of a function), we have

$$x_2 - 2u_2 = 0 \text{ or } u_2 = x_2/2$$

Now using Bellman's principle of optimally, we get

$$f_2(x_2) = (x_2/_2). x_2 - (x_2/2)^2 = x_2^2/4$$

and $f_3(x_3) = \underset{u_3}{\text{Max}} \{u_3 \cdot f_2(x_2)\} = \underset{u_3}{\text{Max}} \{u_3 \cdot (x_2^2/4)\}$

$$= \text{Max} \left\{ u_3 . \frac{(x_3 - u_3)^2}{4} \right\}$$

Again differentiating $f_2(x_2)$ with respect to u_3 and equating to zero,

$$\frac{1}{4} \{u_3 \cdot 2(x_3 - u_3)(-1) + (x_3 - u_3)^2\} = 0$$

$$(x_3 - u_3)(-2u_3 + x_3 - u_3) = 0$$

$$(x_3 - u_3)(x_3 - 3u_3) = 0$$

Now either $u_3 = x_3$ which is travails $u_1 + u_2 + u_3 = x_3$

or $x - 3u_3 = 0$ or $u_3 = x_3/3 = 10/3$. Therefore

$$u_2 = \frac{x_2}{2} = \frac{x_3 - u_3}{2} = \frac{1}{2}\left(10 \frac{10}{3}\right) = \frac{10}{3}$$

$$u_1 = x_2 - u_2 = \frac{20}{3} - \frac{10}{3} = \frac{10}{3}$$

Thus $u_1 = u_2 = u_3 = 10/3$ and hence Max $\{u_1 \cdot u_2 \cdot u_3\}$

$$= (10/3)^3 = 1000/27.$$

Example 5(a):

A company has decided to introduce a product in three phases. Phase 1 will feature making a special offer at a greatly reduce d rate to attract the first-time buyers. Phase 2 will involve intensive advertising to persuade the buyers to continue purchasing at a regular price. Phase 3 will involve a follow up advertising and promotional campaign.

A total of Rs. 5 million has been budgeted for this marketing campaign. If miss the market share captured in Phase 1, fraction f_2 of m is retained in Phase 2, and fraction f_3 of market share in Phase 2 is retained in Phase 3. The expected values of m, f_2 and f_3 at different levels of money expended are given below. How should the money be allocated to the three phases to maximize the final share?

Money		***Effect on Market Share***	
(Rs millions)	***m per cent***	***f2***	***f3***
0	*0*	*0.30*	*0.50*
1	*10*	*0.50*	*0.70*
2	*15*	*0.70*	*0.85*
3	*22*	*0.80*	*0.90*
4	*27*	*0.85*	*0.93*
5	*30*	*0.90*	*0.95*

Solution:

This problem can be treated as a three-stage problem taking each phase as a stage and amount of money spent as the state of the system. Let us adopt the following notations:

s = millions of rupees spent on marketing campaign at a particular stag

x_j = amount of money allocated to phase (stage) j; (j = 1, 2, 3)

$P_j(x_j)$ = market share captured in phase j; (j = 1, 2, 3)

The problem can now be expressed mathematically as:

$$\text{Maximize } Z = p_1(x_1) \cdot p_2(x_2) \cdot p_3(x_3)$$

subject to the constraint

$$x_1 + x_2 + x_3 = 5$$

and $x_1, x_2, x_3 \geq 0$

Using the forward induction approach, the recursive equation calculations are as follows

$$f_j^*(s) = \underset{0 \le x_j \le s}{\text{Max}} \{p_j(s_n) \times f_{j-1}^*(s - x_j)\}$$

Phase 1 (j = 1). If the whole amount is spent in Phase 1, that is $x_1 \le 5$, the optimal return will be

x_1^*	:	0	1	2	3	4	5
$m = f_1^*(x_1)$	:	0	10	15	22	27	30

Phase 2 (j = 2): If the whole amount is spent in Phase 2, then

$$f_2^*(s) = \underset{0 \le x_2 \le s}{\text{Max}} \{f_2(x_2) \times f_1^*(s - x_2)\}$$

$$= \underset{0 \le x_2 \le 5}{\text{Max}} \{f_2(x_2) \times f_1^*(5 - x_2)\}$$

The computations for stages 2 and 3 are shown in Table 1 and 2, respectively.

Table 1 : Phase 2 (j = 2)

	Decision x_2	$f_2(s, x_2) = f_2(x_2) \times f_1^*(s - x_2)$					***Optimal Return***	***Optimal Decision***
States s	*0*	*1*	*2*	*3*	*4*	*5*	$f_2^*(s)$	x_2^*
0	0	–	–	–	–	–	0	0
1	3.0	–	–	–	–	–	3.0	0
2	4.5	5.0	–	–	–	–	5.0	1
3	6.6	7.5	7.0	–	–	–	7.5	1
4	8.1	11.0	10.5	8.0	–	–	11.0	1
5	9.0	13.5	15.4	12.0	8.5	–	15.4	2

Table 2 : Phase 3 (j = 3)

	Decision x_3	$f_3(s, x_3) = f_3(x_3) \times f_2(s - x_3)$					***Optimal Return***	***Optimal Decision***
States *s*	*0*	*1*	*2*	*3*	*4*	*5*	$f_3^*(s)$	x_3^*
5	7.7	7.7	6.37	4.5	2.79	–	7.7	0, 1

In Table 1.7, when $x_3 = 0$,

$s = x_1 + x_2 = 5$,

and Max $f_2^*(s) = 15.4$ for which $x_2 = 2$ and,

therefore, $x_1 = 3$.

But for $x_3 = 1$, $s = x_1 + x_2 = 4$,

and Max $f_2^*(s) = 11.0$ for which $x_2 = 1$ and,

therefore, $x_1 = 3$.

Hence, the optimal policy to obtain a maximum market share of 7.7 per cent can be any one of the following:

(i) spend 3, 2, 0 million rupees in first, second and third phases (stage), respectively, or

(ii) spend 3, 1, 1 million rupees in first, second and third phases (stage), respectively.

Example 5(b):

maximize $Z = 3x_1 + 5x_2 + 4x_3$

subject to $2x_1 + 3x_2 \leq 8$

$$2x_2 + 5x_3 \leq 10$$

$$3x_1 + 2x_2 + 4x_3 \leq 15,$$

$$x_1, x_2, x_3 \geq 0.$$

Solution:

The given problem is of maximization and all the b_i's are non-negative. Converting the inequalities of constraints into equations by introducing slack variables x_4, x_5, x_6. Also the coefficients of slack variables are zero in the objective function. Thus the given problem becomes

$$Z = 3x_1 + 5x_2 + 4x_3 + 0x_4 + 0x_5 + 0x_6$$

subject to $2x_1 + 3x_2 + 0x_3 + x_4 = 8$

$$0x_1 + 2x_2 + 5x_3 + x_5 = 10$$

$$3x_1 + 2x_2 + 4x_3 + x_6 = 15.$$

The starting basic feasible solution is

$x_1 = 0$, $x_2 = 0$, $x_3 = 0$, $x_4 = 8$, $x_5 = 10$, $x_6 = 15$. The solution to the problem using simplex method is given in the following table.

B	c_B	c_j / x_B	3 / Y_1	5 / Y_2	4 / Y_3	0 / Y_4	0 / Y_5	0 / Y_6	Min. ratio x_B/Y_2
Y_4	0	8	2	(3)	0	1	0	0	8/3
Y_5	0	10	0	2	5	0	1	0	10/2 = 5
									←
Y_6	0	15	3	2	4	0	0	1	15/2
$Z=x_B$	$x_B=0$	Δ_j	3	5	4	0	0	0	x_B/Y_3
				↑			↓		
Y_2	5	8/3	2/3	1	0	1/3	0	0	–
Y_5	0	4/3	–4/3	0	(5)	–2/3	1	0	14/15
								←	
Y_6	0	2/3	5/3	0	4	–2/3	0	1	29/12
$Z=c_B$ x_B =40/3		Δ_j	–1/3	0	4	–5/3	0	0	x_B/Y_1
				↑		↓			
Y_2	5	8/3	2/3	1	0	1/3	0	0	4
Y_3	4	14/15	–4/15	0	1	–2/15	1/5	0	–
Y_6	1	89/15	(41/15)	0	0	–2/15	–4/5	1	89/41
									←
$Z = c_B\ x_B$ = 256/45		Δ_j	11/15	0	0	–17/15	–4/5	0	
			↑					↓	
Y_2	5	50/41	0	1	0	15/41	8/41	–10/41	
Y_3	4	62/41	0	0	1	–6/41	5/41	4/41	
Y_1	3	89/41	1	0	0	–2/41	–12/41	15/41	
$Z = c_B\ x_B$ = 765/41		Δ_j	0	0	0	–45/41	–24/41	–11/41	

In the last table all Δ_j's for non-basic variables are negative therefore it gives optimal solution.

∴ Optimal solution is

$$x_1 = \frac{89}{41},\ x_2 = \frac{50}{41},$$

$$x_3 = \frac{62}{41}$$

and Max. $Z = \frac{765}{41}$.

Example 5(c):

Maximize $Z = 3x_1 + 2x_2 + 5x_3$

subject to $x_1 + 2x_2 + x_3 \leq 430$

$$3x_1 + 2x_3 \leq 460$$

$$x_1 + 4x_2 \leq 420,$$

$$x_1, x_2, x_3 \geq 0.$$

Solution:

The given problem is of maximization and all the b_i's are non-negative.

Convert the inequalities of constraints into equations by introducing slack variables x_4, x_5, x_6.

The given problem becomes

$$Z = 3x_1 + 2x_2 + 5x_3 + 0x_4 + 0x_5 + 0x_6$$

		c_j	3	2	5	0	0	0	Min. ratio
B	c_B	x_B	Y_1	Y_2	Y_3	Y_4	Y_5	Y_6	x_B/Y_3
Y_4	0	430	1	2	1	1	0	0	430
Y_5	0	460	3	0	(2)	0	1	0	230
									←
Y_6	0	420	1	4	0	0	0	1	–
$Z=c_B$	$x_B=0$	Δ_j	3	2	5	0	0	0	x_B/Y_2
					↑		↓		
Y_4	0	200	–1/2	(2)	0	1	–1/2	0	100
									←
Y_3	5	230	3/2	0	1	0	1/2	0	–
Y_6	0	420	1	4	0	0	0	1	420/4
Z=1150		Δ_j	–9/2	2	0	0	–5/2	0	
				↑		↓			
Y_1	2	100	–1/4	1	0	1/2	–1/4	0	
Y_3	5	230	3/2	0	1	0	1/2	0	
Y_6	0	20	2	0	0	–2	1	1	
Z=1350	Δ_j	–4	0	0	–1	–2	0		

subject to

$$x_1 + 2x_2 + x_3 + x_4 = 430$$
$$3x_1 + 0x_2 + 2x_3 + x_5 = 460$$
$$x_1 + 4x_2 + 0x_3 + x_6 = 420.$$

The initial basic feasible solution is

$$x_1 = 0, x_2 = 0, x_3 = 0, x_4 = 430,\ x_5 = 460, x_6 = 420.$$

The solution to the problem using simplex algorithm is given in the following table:

Since all Δ_j's are negative or zero therefore the solution is optimal.

Hence the optimal solution is

$$x_1 = 0,\ x_2 = 100,\ x_3 = 230 \text{ and max } Z = 1350.$$

Example 5(d):

Minimize $Z = x_2 - 3x_3 + 2x_5$

subject to

$$3x_2 - x_3 + 2x_5 \leq 7$$
$$-2x_2 + 4x_3 \leq 12$$
$$-4x_2 + 3x_3 + 8x_5 \leq 10,$$
$$x_2, x_3, x_5 \geq 0.$$

Solution:

The given problem is of minimization. Convert it to maximization problem by taking the objective function as $Z' = -Z$.

The objective function becomes

$$\text{Max. } Z' = -Z = -x_2 + 3x_3 - 2x_5.$$

To convert the inequalities of constraints into equations introducing slack variables x_1, x_4 and x_6, we get

$$x_1 + 3x_2 - x_3 + 0x_4 + 2x_5 + 0x_6 = 7$$
$$0x_1 - 2x_2 + 4x_3 + x_4 + 0x_5 + 0x_6 = 12$$
$$0x_1 - 4x_2 + 3x_3 + 0x_4 + 8x_5 + x_6 = 10$$

and the objective function becomes

$$\text{Max. } Z' = 0x_1 - x_2 + 3x_3 + 0x_4 - 2x_5 + 0x_6.$$

The starting basic feasible solution is

$$x_1 = 7,\ x_2 = 0,\ x_3 = 0,\ x_4 = 12,\ x_5 = 0,\ x_6 = 10.$$

The solution to the problem using simplex algorithm is given below:

	c_j	0	–1	3	0	–2	0	Min. ratio	
B	c_B	x_B	Y_1	Y_2	Y_3	Y_4	Y_5	Y_6	x_B/Y_3
Y_1	0	7	1	3	–1	0	2	0	–ive
Y_4	0	12	0	–2	(4)	1	0	0	3 ←
Y_6	0	10	0	–4	3	0	8	1	10/3
$Z'=c_B$	$x_B=0$	Δ_j	0	–1	3 ↑	0 ↓	–2	0	x_B/Y_2
Y_1	0	10	1	(5/2)	0	1/4	2	0	4 ←
Y_3	3	3	0	–1/2	1	1/4	0	0	–ive
Y_6	0	1	0	–5/2	0	–3/4	8	1	–ive
$Z'=9$	Δ_j	0 ↑	1/2 ↓	0	–3/4	–2	0		
Y_2	–1	4	2/5	1	0	1/10	4/5	0	
Y_3	3	5	1/5	0	1	3/10	2/5	0	
Y_6	0	11	1	0	0	–1/2	10	1	
$Z' = 11$	Δ_j	–1/5	0 ↑	0 ↓	–4/5	–12/5	0		

Since all Δ_j's ≤ 0 therefore the solution given by last table is optimal. Hence the optimal solution is

$$x_2 = 4,\ x_3 = 5,\ x_5 = 0 \text{ and min. } Z = -Z' = -11.$$

Example 6:

Solve the following problem by using the two-phase method.

$$\text{Min. } Z = 40x_1 + 24x_2$$

subject to $20x_1 + 50x_2 \geq 4800$

$$80x_1 + 50x_2 \geq 7200,\ x_1,\ x_2 \geq 0.$$

Solution:

The given problem is of minimization. Converting it to maximization by taking the objective function as

$$Z' = -Z.$$

The objective function becomes

$$\text{Max } Z' = -Z = -40x_1 - 24x_2.$$

To change the inequalities of constraint into equations introducing surplus and artificial variables, we get

$$20x_1 + 50x_2 - x_3 + A_1 = 4800$$

$$80x_1 + 50x_1 - x_4 + A_2 = 7200$$

$$x_1, x_2, x_3, x_4, A_1, A_2 \geq 0,$$

x_3, x_4 are surplus variables and A_1, A_2 are artificial variables.

Phase I. Assigning cost 0 to all other variables and cost –1 to artificial variables, the new objective function of auxiliary problem becomes

B	c_B	c_j x_B	0 Y_1	0 Y_2	0 Y_3	0 Y_4	–1 A_1	–1 A_2	Min. ratio x_B/Y_1
A_1	–1	4800	20	50	–1	0	1	0	240
A_2	–1	7200	(80)	50	0	–1	0	1	90 ←
$Z=c_B\ x_B$ = –12000		Δ_j	100 ↑	100	–1	–1	0	0 ↓	
A_1	–1	3000	0	(75/2)	–1	1/4	1	–1/4	80 ←
Y_1	0	90	1	5/8	0	–1/80	0	1/80	144
Z= –2910		Δ_j	0	75/2 ↑	0	1/4	0 ↓	–5/4	
Y_2	0	80	0	1	–2/75	1/50	2/75	–1/50	
Y_1	0	40	1	0	1/60	–1/60	–1/60	–1/60	
Z=0	Δ_j	0	0	0	0	–1	–1		

$$\text{Max } Z' = 0x_1 + 0x_2 + 0x_3 + 0x_4 - A_1 - A_2$$

subject to the constraints given above.

Now applying the simplex method in the usual manner, we have the following table:

Since all $\Delta_j \leq 0$ and no artificial variable appears in the basis therefore an optimum solution to the auxiliary problem has been attained.

Phase II. Now assign the actual costs to the original variables and cost zero to the surplus variables, the objective function becomes

$$\text{Max } Z' = -40x_1 - 24x_2 + 0x_3 + 0x_4.$$

Replace the c_j row values in the final simplex table of phase I by the c_j values of the original objective function.

Also delete the artificial variables from the final simplex table of phase I. Now apply the simplex method in the usual manner to this table.

		c_j	–40	–24	0	0	Min. ratio
B	c_B	x_B	Y_1	Y_2	Y_3	Y_4	x_B/Y_3
x_2	–24	80	0	1	–2/75	1/50	neg.
x_1	–40	40	1	0	(1/60)	–1/60	2400 ←
$Z'=c_B\ x_B= -3520$		Δ_j	0 ↓	0	2/75 ↑	–38/75	
x_2	–24	144	8/5	1	0	–1/50	
x_3	0	2400	60	0	1	–1	
$Z' = -3456$ Δ_j		–8/5	0	0	–12/25		

Since all Δ_j's are negative or zero therefore the solution obtained is optimal.

Hence the optimal solution is $x_1 = 0$, $x_2 = 144$ and max $Z = -Z' = 3456$.

Example 7:

Apply two phase simplex method to solve

$$Min.\ Z = x_1 - 2x_2 - 3x_3$$

subject to $-2x_1 + x_2 + 3x_3 = 2$

$2x_1 + 3x_2 + 4x_3 = 1,$

$x_1, x_2, x_3 \geq 0.$

Solution:

Converting the objective function to maximization form by substituting Z= – Z', we get

$$\text{Max. } Z' = -x_1 + 2x_2 + 3x_3.$$

Introducing the artificial variables A_1 and A_2, the constraint equations become

$$-2x_1 + x_2 + 3x_3 + A_1 = 2$$

$$2x_1 + 3x_2 + 4x_3 + A_2 = 1$$

where $x_1, x_2, x_3, A_1, A_2 \geq 0$.

The new objective function of auxiliary problem is

$$\text{Max } Z' = 0x_1 + 0x_2 + 0x_3 - A_1 - A_2,$$

subject to the constraints mentioned above.

Now apply simplex method in the usual manner to remove artificial variables.

	c_j	0	0	0	–1	–1	Min. ratio	
B	c_B	x_B	Y_1	Y_2	Y_3	A_1	A_2	x_B/Y_3
A_1	–1	2	–2	1	3	1	0	2/3
A_2	–1	1	2	3	(4)	0	1	1/4 ←
Z'= –3	Δ_j	0	4	7 ↑	0	0 ↓		
A_1	–1	5/4	–7/2	–5/4	0	1	–3/4	
Y_3	0	1/4	1/2	3/4	1	0	1/4	
Z'= –5/4	Δ_j	–7/4	–5/4	0	0	–3/4		

Since all Δ_j's are negative or zero, an optimum basic feasible solution to the auxiliary problem has been attained. But *max Z' is negative and artificial variable A_1 appears in the basic solution at a positive level therefore the original L.P.P. does not possess any feasible solution.*

Example 8(a):

Consider the problem of designing electronic devices to carry five power cells, each of which must be located within three electronic systems. If one system's power fails, then it will be powered on an auxiliary basis by the cells of the remaining systems. The probability that any particular system will experience a power failure depends on the number of cells originally assigned to it. Estimated power failure probabilities for a particular system are given below:

Power	***Probability of System Power Failure***		
Cells	***System 1***	***System 2***	***System 3***
1	*0.50*	*0.60*	*0.40*
2	*0.15*	*0.20*	*0.25*
3	*0.04*	*0.10*	*0.10*
4	*0.02*	*0.05*	*0.05*
5	*0.01*	*0.02*	*0.01*

Determine how many power cells should be assigned to each system to maximize the overall system reliability.

Solution:

Let us adopt the following notations:

x_n = number of power cells assigned to stage

$p_n(x_n)$ = probability of power failure for the system n, when it is assigned x_n power cells

$f_n(s)$ = probability that nth and all higher systems will fail, while entering with state s

Here stages correspond to systems and state s is the number of power cells available for allocation at different stages. We shall start from state (power cell) 1. The recursive equations for this problem may be given by

$$f_n(s) = \underset{x_n \le 5}{\text{Min}} \{p_n(x_n) \times f_{n+1}(s - x_n)\}, \; n = 1, 2$$

subject to the constraint

$$x_1 + x_2 + \dots + x_n = 5$$

The dynamic programming calculations are as follows:

Table 1 : Stage 3 (n = 3); $f_3(s) = \min_{x_3 \le 5} \{p_3(x_3)\}$

Decision x_3 / *States, s*	*$p_3(x_3)$* 1	2	3	*Minimum Value* $f_3^*(s)$	*Optimal Decision* x_3^*
1	0.40	–	–	0.40	1
2	0.40	0.25	–	0.25	2
3	0.40	0.25	0.10	0.10	3

Table 2 : Stage 2 (n = 2)

Decision x_2 / *States, s*	*$f_2(s) = p_2(x_2) \times f_s^*(s - x_2)$* 1	2	3	*Minimum Value* $f_2^*(s)$	*Optimal Decision* x_2^*
2	0.24	–	–	0.24	1
3	0.15	0.08	–	0.08	2
4	0.06	0.05	0.04	0.04	3

Table 3 Stage 1 (n = 1)

Section x_1 / *States, s*	$f_1(s) = p_1(x_1) \times f_2^*(s - x_1)$ 1	2	3	*Minimum Value* $f_1^*(s)$	*Optimal Decision* x_1^*
5	0.2	0.012	0.0096	0.0096	3

At stage 1, value of $f_1(s) = 0.0096$ is minimum when $x_1 = 3$ corresponds to system 1. Then for $x_1 = 3$, $x_2 + x_3 = 2$. But at stages 2 and 1, optimal values of x_2 and x_3 are 1 and 1, respectively. Thus optimal solution is: $x_1 = 3$, $x_2 = 1$, $x_3 = 1$ with the smallest probability of total power failure, $f_1(5) = 0.0096$.

Example 8(b):

Use dynamic programming to show that

$$\min_{d_j > 0} = 1$$

and $p_i \geq 0$, for all i

is minimum when $p_1 = p_2 = ... = p_n = 1/n$.

Solution:

The problem is to divide unity into n parts, $p_1, p_2, ... ,p_n$ such that the quantity

$$f_n(1) = \sum_{i=1}^{n} p_i \log p_i$$

is minimum, where $f_n(1)$ denotes the minimum attainable sum of $p_i \log p_i$ (i = 1, 2, ... ,n), when 1 is divided into n parts.

For n = 1 (Stage 1) we have,

$f_1(1) = p_1 \log p_1 = 1 \log 1$, as unity is divided only into $p_1 = 1$ part.

For n = 2 (Stage 2) the unity is to be divided into two parts p_1 and p_2 such that $p_1 + p_2 = 1$. If $p_1 = z$ and $p_2 = 1 - z$, then

$$f_2(1) = \underset{0 < z \leq 1}{\text{Min}} \{p_1 \log p_1 + p_2 \log p_2\}$$

$$= \underset{0 < z \leq 1}{\text{Min}} \{z \log z + (1 - z) \log (1 - z)\}$$

$$= \underset{0 < z \leq 1}{\text{Min}} \{z \log z + f_1(1 - z)\}$$

Similarly, in general, for an n-stage problem, where unity is to be divided into on parts, the recursive equations is

$$f_n(1) = \underset{0 < z \leq 1}{\text{Min}} \{p_1 \log p_1 + p_2 \log p_2 + ... + p_n \log p_n\}$$

$$= \underset{0 < z \leq 1}{\text{Min}} \{z \log z + f_{n-1}(1 - z)\}$$

Solution of recursive equation : The solution to the above recursive equation to get the optimal policy can be obtained with the help of differential calculus.

The minimum value of recursive equation

$$f(z) = z \log (z) + (1 - z) \log (1 - z)$$

is attained at z = 1/2, satisfying the condition $0 < z \leq 1$. Hence, for stage 2, the optimal policy is $p_1 = p_2 = 1/2$ and

$$f_2\ (1) = \frac{1}{2}\log\frac{1}{2}+\left(1-\frac{1}{2}\right)\log\left(1-\frac{1}{2}\right) = 2\ \left(\frac{1}{2}\log\frac{1}{2}\right)$$

Similarly, for stage 3, the minimum value of the recursive equation

$$f_3\ (1) = \underset{0<z\leq 1}{\text{Min}}\ \{z \log z + f_2\ (1-z)$$

$$= \underset{0<z\leq 1}{\text{Min}}\ \left\{z\log z+2\left(\frac{1-z}{2}\right)\log\left(\frac{1-z}{2}\right)\right\}$$

is attained at z = 1/3 satisfying the condition $0 < z \leq 1$. Hence, for stage 3, the optimal policy is, $p_1 = p_2 = 1/3$ and

$$f_3\ (1) = \frac{1}{3}\log\left(\frac{1}{3}\right)+2\left(\frac{1-\frac{1}{3}}{2}\right)\log\left(\frac{1-\frac{1}{3}}{2}\right)$$

$$= \frac{1}{3}\log\left(\frac{1}{3}\right)+2\left\{\frac{1}{3}\log\left(\frac{1}{3}\right)\right\} = 3.\left\{\frac{1}{3}\log\left(\frac{1}{3}\right)\right\}$$

Thus, in general, the optimal policy is

$$p_1 = p_2 = \ldots = p_n = 1/n$$

and $$f_n\ (1) = n\ \left\{\frac{1}{n}\log\left(\frac{1}{n}\right)\right\}$$

The above result can be proved valid for any value of n by using mathematical induction. For n = m + 1, the recursive equation is

$$f_{m+1}\ (1) = \underset{0<z\leq 1}{\text{Min}}\ \{z \log z + f_m\ (1-z)\}$$

Since the minimum of the function

$$F\ (z) = z \log z + \frac{1-z}{m}\log\left(\frac{1-z}{m}\right)$$

is attained for z = 1/(x + 1), therefore using the concept of maximum and minimum, we have

$$f_{m+1}\ (1) = \frac{1}{m+1}\log\left(\frac{1}{m+1}\right)+m\left\{\frac{1}{m+1}\log\left(\frac{1}{m+1}\right)\right\}$$

$$= (m+1)\ \left\{\frac{1}{m+1}\log\left(\frac{1}{m+1}\right)\right\}$$

Hence, the result is true for n = m + 1 also.

Example 8(c):

A salesman located in a city A decided to travel to city B. He knew the distances of alternative routes from city A to city B. He then drew a highway network map as shown in the Fig. 1.3. The city of origin, A, is city 1. The destination city B, is city 10. Other cities through which the salesman will have to pass through are numbered 2 to 9. The arrow representing routes between cities and distances in kilometers are indicated on each route. The salesman's problem is to find the shortest route that covers all the selected cities from A to B.

Solution:

To solve the problem, define problem stages, decision variables, state variables, return function and transition function. For this particular problem, the following definitions will be used to denote various state variables and decision variables.

d_n = decision variables that define the immediate destinations when there are n (n = 1, 2, 3, 4,) stages to go

s_n = state variables describe a specifies city at any stage

D_{sn}, d_n = distance associated with the state variable, sn, and the decision variables, dn for the current nth stage

$f_n(s_n, d_n)$ = minimum total distance for the last n stages, given that salesman is in state s_n and selects d_n as immediate destination

$f_n^*(s_n)$ =optimal path (minimum distance) when the salesman is in state s_n with n more stages to go for reaching the final stage (destination).

We start calculating distances between a pair of cities from destination city 10 (= x_1) and work backwards $x_5 \to x_4 \to x_3 \to x_2 \to x_1$ to find the optimal path. The recursion relationship for this problem can be stated as follows:

$$f_n^*(s_n) = \underset{d_n}{\text{Min}}\left\{d_{s_n, d_n} + f_{n-1}^*(d_n)\right\};\ n = 1, 2, 3, 4$$

where $f_{n-1}^*(d_n)$ is the optimal distance for the previous stages.

Working backward stages from city B to city A, we determine the shortest distance to city B (node 10) in stage 1, from state s_1 = 8 (node 8) and state s_1 = 9 (node 9) in stage 2. Since the distances associated with entering into stage 2 from state s_1 = 8 and s_1 = 9 are $_{D8,}$ 10 = 7 and $D_{9, 10} = 9$, respectively, the optimal value of $f_1^* (s_1)$ is the minimum value between $D_{8, 10}$ and $D_{9, 10}$. The results are shown in Table 1.

Table 1 : Stage 2

	Decision, $d_1 \rightarrow$	$\frac{f_1(s_1, d_1) = D_{s_1, d_1}}{10}$	***Minimum***	***Optimal***
		Distance $f_1^* (s_1)$	***Decision*** d_1	
States, s_1	8	7	7	10
9	9	9	10	

We more backward to stage 3. Suppose that the salesman is at state s_2=5 (node 5). Here he has to decide whether he should go to either d_2 = 8 (node 8) or d_2 = 9 (node 9). For this he must evaluate two sums

$$D_{5,8} + f_1^* (8) = 4 + 7 = 11 \text{ (to state } s_1 = 8)$$

$$D_{5,9} + f_1^* (9) = 8 + 9 = 17 \text{ (to state } s_1 = 9)$$

This distance function for travelling from state s_2 = 5 is the smallest of the set two sums:

$$f_2 (s_2) = \underset{d_2 = 8,9}{\text{Min}} \{11, 17\}$$

$$= 11 \text{ (to state } s_1 = 8)$$

Similarly, the calculation of distance function for travelling from state s_2 = 6 and s_2 = 7 can be completed as follows:

For state, s_2 = 6 $\quad f_2 (6) = \underset{d_2 = 8,9}{\text{Min}} \begin{array}{l} D_{6,8} + f_1^* (8) = 3 + 7 = 10 \\ D_{6,9} + f_1^* (9) = 7 + 9 = 16 \end{array}$

$$= 10 \text{ (to state } s_1 = 8)$$

For state, s_2 = 7 $\quad f_2 (7) = \underset{d_2 = 8,9}{\text{Min}} \begin{array}{l} D_{7,8} + f_1^* (8) = 8 + 7 = 15 \\ D_{7,9} + f_1^* (9) = 4 + 9 = 13 \end{array}$

$$= 13 \text{ (to state } s_1 = 9)$$

These results are entered into the two-stage table as shown in Table 2

Table 2 : Stage 3

Decision, $d_2 \rightarrow f_2(s_2, d_2) = D_{s2,d2} + f_1^*(d_2)$ Optimal				***Minimum***	
		Distance		***Decision***	
		8	***9***	$f_2^*(s_2)$	d_2
States, s_2	5	11	17	11	8
	6	10	16	10	8
	7	15	13	13	9

Continuing the same process for stages 4 and 5, the results are shown in Tables 3 and 4.

Table 3 : Stage 4

Decision, $d_3 \rightarrow f_3(s_3, d_3) = D_{s3,d3} + f_2^*(d_3)$ Optimal					***Minimum***	
			Distance		***Decision***	
		5	***6***	***7***	$f_3^*(s_3)$	d_3
States, s_3	2	18	20	18	18	5 or 7
	3	14	18	17	14	5
	4	17	20	18	17	5

Table 4 : Stage 5

Decision, $d_4 \rightarrow f_4(s_4, d_4) = D_{s4,d4} + f_3^*(d_4)$ Optimal					***Minimum***	
			Distance		***Decision***	
		2	***3***	***4***	$f_4^*(s_4)$	d_4
States, s_4	1	2	20	20	20	3 or 4

The above optimal results at various stages can be summarized as given below:

Entering stages (nodes)

$$\text{Sequence} \begin{cases} 10 \quad 8 \quad 5 \quad 3 \quad 1 \\ 10 \quad 8 \quad 5 \quad 4 \quad 1 \end{cases}$$

$$\text{Distances} \begin{cases} 7 \quad 4 \quad 3 \quad 6 = 20 \\ 7 \quad 4 \quad 6 \quad 3 = 20 \end{cases}$$

From the above, it is clear that there are two alternative shortest routes for this problem, both having a minimum distance of 20 kilometers.

Example 9:

(Optimal sub-division problems) Divide quantity b into n parts so as to maximize their product. Let f_n (b) be the maximum value. Then show that

$$f_1 (b) = b$$

$$f_n (b) = \max_{0 \le z \le b} \{n\, f_{n-1} (b - z)\}$$

Hence find f_n (b) and the division that maximize it.

Solution:

Let x_j be the jth part of the quantity b (j = 1, 2, ... ,n). Then the problem becomes Maximize $f_n (b) = x_1 . x_2 \ldots x_n$

subject to the constraints

$$x_1 + x_2 + \ldots + x_n = b$$

and $\quad x_j > 0;\ j = 1, 2, \ldots ,n$

Here each part x_j (j = 1, 2, ... ,n) of b may be regarded as a stage. Since x_j may assume any positive value satisfying the given condition that $x_1 + x_2 + \ldots + x_n = b$, alternatives at each stage are infinite. Thus x_j's may be considered continuous variables. The recursive equation of the problem for all values of c can be obtained as follows:

For n = 1, the above result becomes f_1 (b) = x or b (initially true).

For n = 2 (*i.e.,* two stage problem), the quantity b is divided into two parts, say $x_1 = z$ and $x_2 = b - z$.

Then

$$f_2 (b) = \text{Max } (x_1, x_2\} = \max_{0 < z \le b} \{z (b - z)\}$$

$$= \max_{0 < z \le b} \{z\, f_1 (b - z)\}, \text{ since } f_1 (b - z) = b - z$$

Similarly, for n = 3, the maximum product of b divided into three parts given the initial choice of z which leaves (b – z) to be further divided into two parts. Denoting the maximum possible product for (b – z) into two parts by f_2 (b – z). Thus, using the principle of optimally, we have

$$f_3(b) = \underset{0<z\le b}{\text{Max}} \{z\, f_2(b-z)\}$$

Continuing in a similar manner, the recursive equation for general value of n is given by

$$f_n(b) = \underset{0<z\le b}{\text{Max}} \{z\, f_{n-1}(b-z)\}$$

Solution to the recursive equation : The solution to recursive equation (1) to get the optimal policy can be obtained with the help of differential calculus.

For n = 2, the function z (b – z) attains its maximum value for z=b/2, satisfying the condition $0 < z \le b$.

Hence, functional equation becomes

$$f_2(b) = \underset{0<z\le b}{\text{Max}} \left\{\frac{b}{2}\cdot\left(b-\frac{b}{2}\right)\right\} = \left(\frac{b}{2}\right)^2$$

Then for n = 2, we have

$$\text{Optimal policy: } \left(\frac{b}{2},\frac{b}{2}\right) \text{ and } f_2(b) = \left(\frac{b}{2}\right)^2$$

For n = 3, the functional equations becomes

$$f_3(b) = \underset{0<z\le b}{\text{Max}} \{z\, f_2(b-x)\} = \underset{0<z\le b}{\text{Max}} \left\{z\left(\frac{v-z}{2}\right)^2\right\}$$

$$= \underset{0<z\le b}{\text{Max}} \left\{z\left(\frac{b-z}{4}\right)^2\right\} = \left(\frac{b}{3}\right)^3$$

The maximum value of $z\left(\frac{b-z}{2}\right)^2$ is attained for $z = \frac{b}{3}$, satisfying the condition $0 < z \le b$ because

$$f_2(b-z) = f_2\left(b-\frac{b}{3}\right) = f_2\left(\frac{2}{3}b\right) = \left\{\frac{1}{2}\left(\frac{2}{3}b\right)\right\}^2 = \left(\frac{1}{3}b\right)^2$$

Thus for n = 3, we have

$$\text{Optimal policy: } \left(\frac{b}{3},\frac{b}{3},\frac{b}{3}\right)$$

and $$f_3(b) = \left(\frac{b}{3}\right)^2$$

Hence, in general, for an n-stage problem, we assume that

$$\text{Optimal policy: } \left(\frac{b}{n}, \frac{b}{n}, \ldots, \frac{b}{n}\right)$$

and $$f_n(b) = \left(\frac{b}{n}\right)^n \text{ form=1, 2, ... ,m.}$$

Now by induction it can also be shown that the result holds good for n = m = 1. The method is discussed below:

For n = m + 1, the functional equation becomes

$$f_{m+1}(b) = \underset{0<z\le b}{\text{Max}} \{z\, f_m(b-z)\}$$

$$= \underset{0<z\le b}{\text{Max}} \left\{ z\left(\frac{b-z}{m}\right)^m \right\}$$

$$= \underset{0<z\le b}{\text{Max}} \left\{ z\frac{(b-z)^m}{(m)^m} \right\} = \left(\frac{b}{m+1}\right)^{m+1}$$

The maximum value of $z\left(\frac{b-z}{m}\right)^m$ is attained for $z = \frac{b}{m+1}$, that is, the result is true for n = m + 1 also.

Hence, the required optimal policy is

$$\left(\frac{b}{n}, \frac{b}{n}, \ldots, \frac{b}{n}\right) \text{ for } f_n(b) = \left(\frac{b}{n}\right)^n$$

Remark: The maximum value $z(b-z)^2$ was obtained by using the concept of maximum and minimum. For example, let

$$f(z) = \left\{\frac{b-z}{2}\right\}^2$$

Then $$\frac{d}{dz}\{f(z)\} = \frac{1}{4}\{2z(b-z)(-1)+(b-z)^2\}$$

But for calculating maximum or minimum of f (z), we have two equate

$$\frac{d}{dz}\{f(z)\} = 0;\ \textit{i.e.,}\ -2z(b-z)+(b-z)^2 = 0$$

This gives z = b/3. Further, the second derivative of f (z), *i.e.*, $\frac{d^2}{dz^2}\{f(z)$ is negative, at z = b/3. Hence the maximum value of f(z) is obtained at z = b/3.

Example 10:

An electronic device consists of four components each of which must function for the system to function. The system reliability can be improved by installing parallel units in one or more components. The reliability of components, R with one, two, or three parallel units, and the corresponding cost, C are given below. The maximum amount available for this device is 100. The problem is to determine the number of parallel units in each component.

Number of	***Components***							
Parallel Units	***1***		***2***		***3***		***4***	
	R	***C***	***R***	***C***	***R***	***C***	***R***	***C***
1	*0.70*	*10*	*0.50*	*20*	*0.70*	*10*	*0.60*	*20*
2	*0.80*	*20*	*0.70*	*40*	*0.90*	*30*	*0.70*	*30*
3	*0.90*	*30*	*0.80*	*50*	*0.95*	*40*	*0.90*	*40*

Solution:

The reliability of the given electronic device is the product of the reliabilities of each of its four components. If R_j and u_j represents the reliability of the component j and units in parallel in component i, then the reliability of whole system consists of n components in series will be:

$$R_1u_1 \times R_2u_2 \times \ldots \times R_nu_n.$$

Since the objective is to maximize the reliability of the system, the problem can be staged as:

$$\text{Maximize } Z = R_1u_1 \times R_2u_2 \times \ldots \times R_nu_n$$

subject to the constraint

$$c_1u_1 + c_2u_2 + \ldots + c_nu_n \leq C$$

where $c_j\, u_j$ = cost of component j

C = total capital available

Since the electronic device consists of n = 4 components, for solving this problem consider each component as a stage. The state at any stage will be the capital to be allocated. Let us adopt the following notations.

x_j = capital allocated to stage j, through first stage inclusive.

$f_j(x_j)$ = return when available capital cj is allocated optimally over n stages (components)

$R_j u_j$ = reliability of component j (j = 1, 2, 3, 4)

Now the recursive equation can be expressed as:

$$f_j(x_j) = \underset{0 < c_j u_j < x_j}{\text{Max}} [\{R_1 u_1 \times R_2 u_2 \times \ldots \times R_n u_n\}.\ f_{j-1}(x_j - c_j u_j];$$

j = 1, 2, 3, 4.

The electronic device in this problem will consist of at least one unit in each component. Thus, the range of investment, xj (j = 1, 2, 3, 4) in each case will be as follow:

$$c_{11} \le x_1 \le C - c_{21} - c_{31} - c_{41} \text{ or } 10 \le x_1 \le 50$$
$$c_{11} + c_{12} \le x_2 \le C - c_{31} - c_{41} \text{ or } 30 \le x_2 \le 70$$
$$c_{11} + c_{12} + c_{13} \le x_3 \le C - c_{41} \text{ or } 40 \le x_3 \le 80$$
$$c_{11} + c_{12} + c_{12} + c_{14} \le x_4 \le C \text{ or } 60 \le x_4 \le 100$$

The computation of return at each stage using the forward induction approach, is shown in the following tables.

Table 1 : Stage 1 (j = 1)

Decision	*$f_1(x_1) = R_1 u_1$* *u_1*			*Optimal Return*	*Optimal Decision*
States	$u_1 = 1$ R = 0.70; C = 10	$u_1 = 2$ R = 0.80; C = 20	$u_1 = 3$ R = 0.90; C = 30	*$f_1^*(x_1)$*	*u_1^**
10	0.70	–	–	0.70	1
20	0.70	0.80	–	0.80	2
30	0.70	0.80	0.90	0.90	3
40	0.70	0.80	0.90	0.90	3
50	0.70	0.80	0.90	0.90	3

Table 2 : Stage 2 (j = 2)

Decision u_2 / States	$f_2(x_2) = R_2u_2 \times f_1^*(x_2-c_2x_2)$			Optimal Return	Optimal Decision
	$u_2 = 1$ R = 0.50; C = 20	$u_2 = 2$ R = 0.70; C = 40	$u_2 = 3$ R = 0.80; C = 50	$f_2^*(x_2)$	u_2^*
30	0.5×0.7 = 0.35	–	–	0.35	1
40	0.5×0.8 = 0.40	–	–	0.40	1
50	0.5×0.9 = 0.45	0.7×0.7 = 0.49	–	0.49	2
60	0.5×0.9 = 0.45	0.7×0.8 = 0.56	0.8×0.7. = 0.56	0.56	2, 3
70	0.5×0.9 = 0.45	0.7×0.9 = 0.63	0.8×0.8 = 0.64	0.64	3

Table 3 : Stage 3 (j = 3)

Decision u_3 / States	$f_3(x_3) = R_3u_3 \times f_2^*(x_3-c_3x_3)$			Optimal Return	Optimal Decision
	$u_3 = 1$ R = 0.70; C = 10	$u_3 = 3$ R = 0.90; C = 40	$u_3 = 3$ R = 0.90; C = 40	$f_3^*(x_3)$	u_3^* x_3
40	0.7×0.35=0.245	–	–	0.245	1
50	0.7×0.40=0.280	–	–	0.280	1
60	0.7×0.49=0.343	0.9×0.35=0.315	–	0.343	1
70	0.7×0.56=0.392	0.9×0.40=0.360	0.95×0.35=0.3325	0.392	1
80	0.7×0.64=0.448	0.9×0.49=0.441	0.95×0.40=0.3800	0.448	1

Table 4 : Stage 4 (j = 4)

Decision u_4 / States	$f_4(x_4) = R_4u_4 \times f_3^*(x_4 - c_4u_4)$			Optimal Return	Optimal Decision
	$u_4 = 1$ R = 0.60; C = 20	$u_4 = 2$ R = 0.70; C = 30	$u_4 = 3$ R = 0.90; C = 40	f4* (x4)	u_4^* x_4
60	0.6×0.245=0.147	–	–	0.147	1
70	0.6×0.280=0.168	0.7×0.245=0.171	–	0.171	2
80	0.6×0.343=0.205	0.7×0.280=0.196	0.9×0.245=0.220	0.220	3
90	0.6×0.392=0.235	0.7×0.343=0.240	0.9×0.280=0.252	0.252	3
100.	0.6×0.448=0.268	0.7×0.392=0.274	0.9×0.343=0.308	0.308	3

In Table 1.11, at stage 4 the value of return function f_4 (x_4) is maximum, *i.e.,* 0.308 at x_4 = 100, and u_4 = 3. This makes x_3 = 100 – 40 = 60. In Table 1.10, x_3 = 60 corresponds to u_3 = 1, and we are left with x_2 = 60 – 10 = 50. In Table 1.9, x_2 = 50 corresponds to u_2 = 2, and we are left with x_1 = 50 – 40 = 10. In Table 1.8, x_1 = 10 corresponds to u_1 = 1.

Hence for maximum reliability, the device must have 1, 2, 1 and 3 units in components 1, 2, 3 and 4, respectively to attain a maximum reliability of 0.308 or 30.8 per cent.

Example 11(a):

Use dynamic programming to solve the following problem:

Minimize $Z = y_1^2 + y_2^2 + y_3^2$

subject to the constraint

$$y_1 + y_2 + y_3 = 10$$

and $y_1, y_2, y_3 \geq 0$

Solution:

Let the state variables be s_1, s_2 and s_3 such that

$$s_3 = y_1 + y_2 + y_3 = 15$$

$$s_2 = y_1 + y_2 = s_3 - y_3$$

$$s_1 = y_1 = s_2 - y_2$$

The recursive equations can now be written as:

$$f_3(s_3) = \min_{y_3} \{y_3^2 + f_2(s_2)\}$$

$$f_2(s_2) = \min_{y_2} \{y_2^2 + f_1(s_1)\}$$

$$f_1(s_1) = \min_{y_1} \{y_1^2\} = y_1^2 = (s_2 - y_2)^2$$

The recursive equation f_2 (s_2) can also be expressed as:

$$f_2(s_2) = \min_{y_2} \{y_2^2 (s_2 - y_2)^2\};\ f_1(s_1) = (s_2 - y_2)^2$$

Now by using the concept of maxima and minima in differential calculus, the minimum value of f_2 (s_2) can be obtained as explained below:

Differentiating $f_2(s_2)$ with respect to y_2 and equating to zero, we get

$$2y_2 - 2(s_2 - y_2) = \text{ or } y_2 = s_2/2$$

Thus $$f_1(s_2) = \left(\frac{s_2}{2}\right)^2 + \left\{s_2 - \frac{s_2}{2}\right\}^2 = \frac{s_2^2}{2}$$

$$f_3(s_3) = \underset{y_3}{\text{Min}} \; [y_3^2 + f_2(s_2)\} = \underset{y_3}{\text{Min}} \left\{y_3^2 + \frac{s_2^2}{2}\right\}$$

$$= \underset{y_3}{\text{Min}} \left\{y_3^2 + \frac{(s_2 - y_3)^2}{2}\right\}$$

Differentiating $f_3(s_3)$ with respect to y_3 and equating to zero to get its minimum value,

$$2y_3 - (s_3 - y_3) = 0 \text{ or } y_3 - s_3/3$$

Thus $$f_3(s_3) = \left(\frac{s_3}{3}\right)^2 + \frac{1}{2}\left(s_3 - \frac{s_3}{3}\right)^3 = \frac{s_3^2}{3}.$$

But for minimum value of $f_3(s_3)$ we must have $y_1 + y_2 + y_3 = 10$. Therefore

$$f_3(s_3) = (10)2/3 = 100/3, \text{ and}$$

$$y_3 = s_3/3 = 10/3$$

$$y_2 = s_2/2 = (s_3 - y_3)/2 = \{10 - (10/3)\}/2 = 10/3$$

$$y_1 = s_1/2 = (s_3 - y_3) - y_2 = \{10 - (10/3)\{ - 10/3 = 10/3$$

Hence, the minimum value of $y_1^2 + y_2^2 + y_3^2$ is 100/3

for $y_1 = y_2 = y_3 = 10/3$.

Example 11(b):

Suppose there are n machines which can perform two jobs. If x of them do the first job, then they produce goods worth g(x) = 3x, and if y of them perform the second job, then they produce goods worth h(y) = 2.5y. Machines are subject to depreciation so that after performing the first job only a(x) = x/3 machines remain available and after perforing the second job b(y) = 2/3 machines remain available in the beginning of the second year. The process is repeated with remaining machines. Obtain the maximum total return after three years and also find the optimal policy in each year.

Solution:

Let us define the following notations

n = stages, each year being viewed as a stage; n = 1, 2, 3

x_n = number of machines devoted to job 1 in the year n

y_n = number of machines devoted to job 2 in the year n

s = state variable, the number of machines at any stage (year)

f_j (s) = maximum return function when initial available machines are s with n more stages (years) to go

Stage 1 (n = 1) In this problem, there are three stages (years). Using the backward induction approach, *i.e.,* n = 1 (third year), we have s = s_3 number of machines at the beginning of the year. Then the return function can be expressed as:

$$f_1(s_3) = \max_{x_3, y_3} \{3x_3 + 2.5y_3\}$$

subject to the constraint

$$x_3 + y_3 \le s_3 \qquad \text{and} \qquad x_3, y_3 \ge 0$$

where x_3, y_3 = number of machines devoted to jobs 1 and 2, respectively in the third year.

Since function $f_1(s_3)$ is a linear function in x_3 and y_3, its maximum can be obtained by using the concept of extreme point solutions of LP problem. As shown in Fig. 1.4, the maximum value of return function occurs at B (s_3, 0), where

$$f_1(s_3) = 3x_3 + 2.5y_3$$

$$= 3s_3 + 2.5 \times 0 = 3s_3.$$

Hence optimal decision at this stage is: $x_3^* = s_3$, $y_3^* = 0$

and $f_1^*(s_3) = 3s_3$.

Stage 2 (n = 2) : Consider now the second year as the second stage. The return function at this stage is

$$f_2(s_2) = \max \begin{Bmatrix} \text{Immediate return} & \text{Maximum known return} \\ \text{from stage +} & \text{from stage 1} \end{Bmatrix}$$

$$= \max_{x_2, y_2} \left\{(3x_2 + 2.5y_2) + f_1^*\left(\frac{x_2}{3} + \frac{2y_2}{3}\right)\right\} y_2$$

subject to the constraint

$$x_2 + y_2 \le s_2 \quad \text{and} \quad x_2, y_2 \ge 0$$

where x_2 and y_2 = number of machines devoted to jobs 1 and 2, respectively in the second year.

$x_2/3$ and $2y_2/3$ = machines which will be available at the beginning of the next year.

By the definition of $f_1^*(s_3) = 3s_3$ (Stage 1) the return function of this stage becomes

$$f_2(s_2) = \underset{x_2, y_2}{\text{Max}} \left\{3x_2 + 2.5.y_2 + 3\left(\frac{x_2}{3} + \frac{2y_2}{3}\right)\right\}$$

$$= \underset{x_2, y_2}{\text{Max}} \{4x_2 + 4.5y_2\}$$

This is again a linear function in x_2 and y_2, and its maximum value occurs at corner point A $(0, s_2)$.

The maximum value of the return function at A $(0, s_2)$ is: $f_2(s_2) = 4 \times 0 + 4.5 \times s_2 = 4.5s_2$. Hence, the optimal decision at this stage is: $x_2^* = 0$, $y_2^* = s_2$ and $f_2^*(s_2) = 4.5s_2$

Stage 3 (n = 3) Consider now the first year as the third stage. The return function at this stage is

$$f_3(s_1) = \begin{Bmatrix} \text{Im mediate return} & \text{Maximum known return} \\ \text{from stage } 3+ & \text{from stage } 2 \end{Bmatrix}$$

$$= \underset{x_1, y_1}{\text{Max}} \left\{(3x_1 + 2.5y_1) + f_2^*\left(\frac{x_1}{3} + \frac{2y_1}{3}\right)\right\}$$

$$= \underset{x_1, y_1}{\text{Max}} \left\{3x_1 + 2.5y_1 + 4.5\left(\frac{x_1}{3} + \frac{2y_1}{3}\right)\right\} \quad f_2(s_2) = 4.5s$$

$$= \underset{x_1, y_1}{\text{Max}} \{4.5x_1 + 5.5y_1\}$$

subject to the constraint

$$x_1 + y_1 \le s_1$$

and $x_1, y_1 \ge 0$

This return function is also a linear function in x_1 and y_1, and its maximum value occurs at corner point A (0, s1) as shown in Fig. 1.6. The maximum value of return function at A (0, s1) is

$$f_3(s_1) = 4.5 \times 0 + 5.5 \times s_1 = 5.5s_1$$

Hence the optimal decision at this stage is

$$x_1^* = 0, y_1^* = s_1 \text{ (= m machines, say) and } f_3^*(s_1) = 5.5m$$

The values of x_2, x_3 and y_2, y_2 can also be obtained in terms of m as follows:

$$y_2 = s_2 = 2y_1/3 = 2m/3 \text{ and } x_2 = 0$$

$$x_3 = s_3 = 2y_2/3 = 2/3\ (2m/3) = 4m/9, \text{ and } y_3 = 0$$

Table 1 summarizes the optimal decision at each stage (year).

Table 1 : Summary of the Optimal Decisions

Stage	*Values of Decision variable*
3 (First year)	$x_1 = 0,\ y_1 = m$
2 (Second year)	$x_2 = 0,\ y_2 = 2m/3$
1 (Third year)	$x_3 = 4x/9,\ y_3 = 0$

The maximum possible return is 5.5m.

Example 11(c):

A man is engaged in buying and selling identical items, he operates from a warehouse having a capacity of 500 items. Each month he can sell any quantity that he chooses up to the stock at the beginning of the month. Each month, he can buy as much as he wishes for delivery at the end of the month so long as his stock does not exceed 500 items. For the next four months he has the following error free forecasts of cost and sales prices:

Month n	:	*1*	*2*	*3*	*4*
Cost, c_n	:	*27*	*24*	*26*	*28*
Sales price, p_n	:	*28*	*25*	*25*	*27*

If he currently has a stock of 200 units, what quantities should he and buy in the next four months? Find the solution using dynamic programming.

Solution:

Let us adopt the following notations:

x_n = amount to be purchased during month, n

y_n = amount to be sold during month, n

s_n = starting inventory for each month, n (state variable)

p_n = sales price in month, n

c_n = purchase price in month, n

$f_n(s_n)$ = optimal total return from the next n stages if stage n begins with inventory level sn.

To solve this problem we will use backward induction approach, *i.e.*, month 4 be stage 1 and month 1 be stage 4 so that the stage number measures the future time remaining.

The state of the warehouse is the starting inventory s_n for each month. The state transformation therefore, given by

$$s_{n-1} = \text{Inventory at start of month } n - 1$$
$$= s_n + x_n - y_n$$

Also, since selling precedes buying, the man cannot sell more than he owns at the start of the month: that is, $y_n \le s_n$

The ending inventory for the month cannot exceed the warehouse capacity. Therefore

$$x_n + x_n - y_n \le 500$$

or Purchase order, $x_n \le 500 - s_n + y_n$

The return from each month's buying minus selling is the net revenue, *i.e.*,

Revenue, $r_n = y_n P_n - x_n c_n$

Let $f_n(s_n)$ represent the maximum return from the stage n, if it begins with inventory level s_n. Then recursive equation becomes

$$f_n^*(s_n) = \text{Max}\,\{y_n p_n - x_n c_n + f_{n-1}^*(s_n + x_n - y_n)\}$$

where $0 \le y_n \le s_n$

$0 \le x_n \le 500 - s_n + y_n$

Stage 1 (n = 1) : Since this sub-problem has no future to follow it, the return function is given by

$f_1^* (s_1) = \text{Max } (27y_1 - 28x_1)$

where $0 \le y_1 \le s_1$

$0 \le x_1 \le 500 - s_1 + y_1$

This represent a linear programming problem with two variables. The solution of it can be obtained graphically. The value of the objective function increases with the value of y_1 and decreases with value of x1. Therefore, the optimal solution is: $x_1 = 0, y_1 = s_1$ with the maximum return

$$f_1^* (s_1) = p_1y_1 = 27s_1$$

and $s_1 = s_2 + x_2 - y_2$

Stage 2 (n = 2) : $f_2^* (s_2) = \text{Max } \{25y_2 - 26x_2 + f_1^* (s_2 + x_2 - y_2)\}$

where $0 \le y_2 \le s_2$

$0 \le x_2 \le 500 = s_2 + y_2$

Since $f_1^* (s_1) = 27s_1$, substituting this value in the expression for f_2 (s_2), we have

$$f_2^* (s_2) = \text{Max } \{25y_2 - 26x_2 + 27 (s_2 + x_2 - y_2)\}$$

$$= \text{Max } \{27s_2 + x_2 - 2y_2\}$$

The optimal solution at this stage can be obtained by comparing the following alternative solutions (corners of the feasible solution space).

(i) $x_2 = 0, \quad y_2 = 0, \quad f_2 (s2) = 27s_2$

(ii) $x_2 = 0, \quad y_2 = s_2, \quad f_2 (s2) = 25s_2$

(iii) $x_2 = 500, \quad y_2 = s_2, \quad f_2 (s_2) = 25s_2 + 500$

(iv) $x_2 = 500 - s_2, \quad y_2 = 0, \quad f_2 (s_2) = 26s_2 + 500$

Hence for any value of $s_2 \le 500$, the optimal return is

$$f_2^* (s_2) = 26s_2 + 500; \; s_2 = s_3 + x_3 - y_3$$

for $x_2 = 500 - s_2$ and $y_2 = 0$

Stage 3 (n = 3) : $f_3^* (s_3) = \text{Max } \{25y_3 - 24x_3 + f_3^* (s_3 + x_3 - y_3)\}$

$= \text{Max } \{25y_3 - 24x_3 + 26 (s_3 + x_3 - y_3) + 500\}$

$= \text{Max } \{26s_3 + 2x_3 - y_3 + 500\}$ the optimal value of $f_3 (s_3)$ can be obtained by comparing the following alternative solution:

(i) $x_3 = 0, \quad y_3 = 0, \quad f_3 (s_3 = 26s_3 + 500$

(ii) $x_3 = 0$, $y_3 = s_2$, f_3 ($s_3 = 25s_3 + 500$

(iii) $x_3 = 500$, $y_3 = s_2$, $f_3(s_3) = 25s_3 + 1{,}500$

(iv) $x_3 = 500 - s_3$, $y_3 = 0$, $f_2(s_2) = 24s_3 + 1{,}500$

Hence for any value of $s_3 \leq 500$, the optimal return is

$$f_3^*(s_3) = 25s_3 + 1{,}500;\ s_3 = s_4 + x_4 - y_4$$

for $x_3 = 500$ and $y_2 = s_3$

Stage 4 (n = 4) : $f_4^*(s_4) = \text{Max}\ \{28y_4 - 27x_4 + f_3^*(s_4 + x_4 - y_4)\}$

$= \text{Max}\ \{28y_4 - 27x_4 + 25(s_4 + x_4 - y_4) + 1{,}500\}$

$= \text{Max}\ \{25s_4 - 2x_4 + 3y_4 + 1{,}500\}$

Again comparing the four alternative solutions, the optimal return is obtained at $x_4 = 0$ and $y_4 = s_4$, and

$$f_4^*(s_4) = 28s_4 + 1{,}500$$

However, it is given that, $s_4 = 200$, $x_4 = 0$; $y_4 = 200$. Then,

$s_3 = s_4 - y_4 + x_4$

$= 200 - 200 + 0 = 0;$

$x_3 = 500$, $y_3 = 0$

$s_2 = s_3 - y_3 + x_3$

$= 0 - 0 + 500 = 500;$

$x_2 = 0$, $y_2 = 0$

$s_1 = s_2 - y_2 + x_2$

$= 500 - 0 + 0 = 500;$

$x_1 = 0$, $y_2 = 0$

Thus, the required solution is as follows:

Month	:	1	2	3	4
Purchase	:	0	500	0	0
Sale	:	200	0	0	500

and maximum possible return = 28 (200) + 1,500 = Rs. 7,100.

Example 11(d):

A company has five salesmen, who have to be allocated to four marketing zones. The return (or profit) from each zone depends upon the number of salesmen working in that zone. The expected returns for different number of salesmen in different zones, as estimated from the past records are given below. Determine the optimal allocation policy.

Number of Salesmen	*Marketing Zones* 1	2	3
0	45	30	35
1	58	45	45
2	70	60	52
3	82	70	64
4	83	79	72
5	101	90	82

Solution:

This problem can be treated as a three-stage problem taking each marketing zone as a stage the number of salesmen allocated to a particular zone as a state variable. Let us adopt the following notations:

s = total number of salesmen available

x_j = number of salesmen allocated to marketing zone j (j = 1, 2, 3)

$p_j(x_j)$ = return from zone j when x_j salesmen are allocated, j = 1, 2, 3

Now the given problem can be mathematically stated as:

$$\text{Maximize } Z = p_1(x_1) + p_2(x_2) + p_3(x_3)$$

subject to the constraint

$$x_1 + x_2 + x_3 \leq 5$$

and $x_1, x_2, x_3 \geq 0$ and integers.

To obtain the recurrence relation, let there be s salesmen available to be allocated to all marketing zones. Then $f_j(s)$ represents the optimal allocation of salesmen to marketing zone j and x_j the initial location to marketing zone j. Then the general recursive equation becomes:

$$f_j(s) = \underset{0 \leq x_j \leq s}{\text{Max}} \{p_j(x_j) + f^*_{j+1}(s - x_j)\};\ j = 1, 2, 3$$

Using backward induction, we start by optimizing the last stage, stage 3 – marketing zone 3. Computations at each stage are shown below:

Stage 2 (j = 3)

s	:	0	1	2	3	4	5
f_3^* (s)	:	35	45	52	64	72	82
x_3^*	:	0	1	2	3	4	5

Stage 3 (j = 2) For this stage recursive equation will become

$$f_2(s) = \underset{0 \le x_2 \le s}{\text{Max}} \{p_2(x_2) + f_3^*(s - x)\}$$

The computations for stages 2 and 1 are shown in Tables 1 and 2, respectively.

Table 1 : Computations for Stage 2

Decision x_2	$f_2^*(s) = p_2(x_2) + f_2(s - x_2)$						*Optimal Return*	*Optimal Decision*
States, s	*0*	*1*	*2*	*3*	*4*	*5*		x_2^*
0	65	–	–	–	–	–	65	0
1	75	80	–	–	–	–	80	1
2	82	90	95	–	–	–	95	2
3	94	97	105	105	–	–	105	2, 3
4	102	109	112	115	114	–	115	3
5	112	117	124	112	124	125	125	5

Stage 1 (j = 1)

Table 2 : Computations for Stage 1

Decision x_1	$f_1^*(s) = p_1(x_1) + f_2^*(s - x_1)$						*Optimal Return*	*Optimal Decision*
States, s	*0*	*1*	*2*	*3*	*4*	*5*		x_1^*
5	170	172	175	177	173	166	177	3

In Table 2, the maximum return, Rs 177, corresponds to $x_1 = 3$ salesmen, which leaves, s = 5 – 3 = 2 salesmen for two other stages. From Table 1, for s = 2, we have $x_2 = 2$. This leaves s= 2 – 2 = 0, *i.e.*, no salesman for stage 1 (marketing zone 3). Hence, the optimal solution so obtained is:

Stage (Marketing zone)	*Value of Decision Variable (Salesmen)*
3 (Zone 3)	0
2 (Zone 2)	2
1 (Zone 1)	3.

Example 11(e):

The owner of a chain of four grocery stores has purchased six crates of fresh strawberries. The estimated probability distribution of potential sales of the strawberries before spoilage differs among the four stores. The following table gives the estimated total expected profit at each store, when it is allocated various number of crates.

Number of Crates	*Stores*			
	1	*2*	*3*	*4*
0	*0*	*0*	*0*	*0*
1	*4*	*2*	*6*	*2*
2	*6*	*4*	*8*	*3*
3	*7*	*6*	*8*	*4*
4	*7*	*8*	*8*	*4*
5	*7*	*9*	*8*	*4*
6	*7*	*10*	*8*	*4*

For administrative reasons, the owner does not wish to split crates between stores. However, he is willing to distribute zero crates to any of his stores. Find the allocation of six crates to four stores so as to maximize the expected profit.

Solution:

To solve this problem using dynamic programming, we view each store as a stage. We define the following notation.

x_j = number of crates allocated to stage (store) j, (j = 1, 2, 3, 4)

$p_j(x_j)$ = expected profit from allocation of x_j crates to stage (store) j, (j = 1, 2, 3, 4)

Now the problem can be stated mathematically as:

$$\text{Maximize } Z = p_1 (x_1) + p_2 (x_2) + p_3 (x_3) + p_4 (x_4)$$

subject to the constraint

$$x_1 + x_2 + x_3 = x_4 = 6 \text{ and } x_1, x_2, x_3, x_4 \geq 0 \text{ and integers.}$$

To obtain the recurrence relation, let there be s crates available to be distributed among stores (stages) and x_j the initial allocation to store j.

Then f_j (s) represents the profit associated with the optimal allocation of crates to store (stage) 1 through 4 both inclusive. Thus, the general recurrence equation will become

$$f_j^* (s) = \underset{0 \leq x_j \leq s}{\text{Max}} \{p_j (x_j) + f_{j+1}^* (s - x_j)\}; \; j = 1, 2, 3, 4$$

Using backward induction, we start by optimizing the last stage; stage 4, *i.e.*, store 4. The computations at each stage are shown below:

Stage 4 (j = 4)

s	:	0	1	2	3	4	5	6
f_4^* (s)	:	0	2	3	4	4	4	4
x_4^*	:	0	1	2	3	4	5	6

Stage 3 (j = 3)

Table 1 : Computations for Stage 3

Decision	$f_3^* (s) = p_3 (x_3) + f_4^* (s - x_3)$							***OptimalOptimal***	
x_3								***ReturnDecision***	
States, s	*0*	*1*	*2*	*3*	*4*	*5*	*6*	$f_3^*(s)$	x_3^*
0	0	–	–	–	–	–	–	0	0
1	2	6	–	–	–	–	–	6	1
2	3	8	8	–	–	–	–	8	1, 2
3	4	9	10	8	–	–	–	10	2
4	4	10	11	10	8	–	–	11	2
5	4	10	12	11	10	8	–	12	2
6	4	10	12	12	11	1	8	12	2, 3

Stage 2 (j = 2)

Table 2 : Computations for Stage 2

Decision x_2 / *Stages, s*	$f_2^*(s) = p_2(x_2) + f_3^*(s - x_2)$ *0*	*1*	*2*	*3*	*4*	*5*	*6*	*Optimal Return* $f_2^*(s)$	*Optimal Decision* x_2^*
0	0	–	–	–	–	–	–	0	0
1	6	2	–	–	–	–	–	6	1
2	8	8	4	–	–	–	–	8	0, 1
3	10	10	10	6	–	–	–	10	0, 1, 2
4	11	12	12	12	8	–	–	12	1, 2, 3
5	12	13	14	14	14	9	–	14	2, 3, 4
6	12	14	15	16	16	10	15	16	3, 4

Stage 1 (j = 1)

Table 3 : Computations for Stage 1

Decision x_1 / *Stages, s*	$f_1^*(s) = p_1(x_1) + f_2^*(s - x_1)$ *0*	*1*	*2*	*3*	*4*	*5*	*6*	*Optimal Return* $f_1^*(s)$	*Optimal Decision* x_1^*
6	16	18	18	17	15	13	7	18	1, 2

In Table 3, the maximum return Rs 18 corresponds to the decision, $x_1 = 1$ and $x_1 = 2$ of allocating crates to store 1 (stage 1). When $x_1 = 1$, there remain s = 6 – 1 = 5 crates to be allocated among three other stores. At stage 2, 5 crates (state) corresponds to $x_2 = 2$, 3 or 4. If $x_2 = 2$, then there remain s = 6 – 3 = 3 crates to be allocated to stage 3 and 4. At stage 3, 3 crates corresponds to $x_3 = 2$ and x4 = 1 at stage 4. Hence, one optimal allocation is:

$x_1 = 1, \; x_2 = 2,$

$x_3 = 2, \; x_4 = 1$

$x_1 = 2, \; x_2 = 1,$

$x_3 = 2, \; x_4 = 1.$

Other alternative allocations of crates among stores are as follows to obtain the maximum profit of Rs 18.

Store 1, x_1	:	1	1	1	1	2	2	2	2
Store 2, x_2	:	2	3	3	4	1	2	2	3
Store 3, x_3	:	2	1	2	1	2	1	2	1
Store 4, x_4	:	1	1	0	0	1	1	0	0

Example 11(f):

Find the minimum value of

$$Z = y_1^2 + y_2^2 + \dots + y_n^2$$

subject to the constraint

$$y_1 \cdot y_2 \dots y_n = c \ (c \geq 0)$$

and $y_j \geq 0;\ j = 1, 2, \dots, n.$

Solution:

Let f_n (c) be the minimum attainable value of Z when c is divided into n factors.

For n = 1, we have y_1 = c and therefore

$$f_1(c) = \underset{y_1 = c}{\text{Min}} \{y_1^2\} = c_2$$

For n = 2, let y_1 = x and y_2 = c/x. Then

$$f_2(c) = \underset{0 < x \leq c}{\text{Min}} \{x_2 + f_1 (c/x)^2\}$$

$$= \underset{0 < x \leq c}{\text{Min}} \{x_2 + f_1 (c/x)^2\}$$

Since f_1 (c) = c_2, we have f_1 (c/x) = $(c/x)^2$

For n = 3, let y_1 = x, $y_2 \cdot y_3$ = c/x. Then,

$$f_3(c) = \underset{0 < x \leq c}{\text{Min}} \{y_1^2 + y_2^2 + y_3^2\} = \underset{0 < x \leq c}{\text{Min}} \{x_2 + f_2 (c/x)\}$$

Proceeding in the same way and using the principle of optimally, the recursive equation will become

$$f_n(c) = \underset{0 < x \leq c}{\text{Min}} \{x_2 + f_{n-1} (c/x)\}; \text{ for all n}$$

Solution of recursive equation : The solution to the recursive equation to get the optimal policy can be obtained with the help of differential calculus. Let f (x) = $x_2 + (c/x)^2$. Then

$$\frac{df}{dx} = 2x - \frac{2c^2}{x^2} = 0$$

which gives $x = (c)^{1/2}$, therefore, $y_1 = (c)^{1/2}$ and $y_2 = c/x = (c)^{1/2}$.

Since the second derivative of f (x) with respect to x is positive, f (x) is minimum. Also

$$f_2(c) = \underset{0 < x \le c}{\text{Min}} \{x_2 + (c/x)^2\}$$

$$= (c^{1/2})^2 + \left(\frac{c}{c^{1/2}}\right)^2 = 2c$$

Thus $\quad f_2(c/x) = 2(c/x)$

Hence the optimal policy is $(c^{1/2}, c^{1/2})$ and $f_2(c) = 2c$. Again

$$f_3(c) = \underset{0 < x \le c}{\text{Min}} \{x_2 + f_2(c/x)\} = \underset{0 < x \le c}{\text{Min}} \{x_2 + 2(c/x)\}$$

$$= (c^{1/3})^2 + 2\left(\frac{c}{c^{1/3}}\right) = 3c^{2/3}$$

since minimum of $f(x) = x_2 + 2c/x$ was obtained at $x = c^{1/3}$. Hence the optimal policy is: $\{c^{1/3}, c^{1/3}, c^{1/3}\}$ and $f_3(c) = 3c^{1/3}$.

Continuing in this manner, the optimal policy for an n stage problem will be

$$\{c^{1/n}, c^{1/n}, \ldots, c^{1/n}\} \text{ and } f_n(c) = nc^{2/n}$$

Example 12:

Use dynamic programming to solve the following linear programming problem.

Maximize $Z = 3x_1 = 5x_2$

subject to the constraints

$$x_1 \le 4,\ x_2 \le 6,\ 3x_1 + 2x_2 \le 18$$

and $x_1, x_2 \ge 0$

Solution:

This linear programming problem can be considered as a two-stage, three-stage problem because there are two decision variables and three constraints with available resources.

The optimal value of f_1 (b_1, b_2, b_3) at the first stage is given by

$$f_1(b_1, b_2, b_3) = \underset{0 \le x_1 \le b}{\text{Max}} \{3x_1\}$$

where $b_1 = 4$, $b_2 = 6$ and $b_3 = 18$.

The feasible value of x_1 is a non-negative value which satisfies all the given constraints $x_1 \le b_1$ (= 4), $3x_1 \le b_3$ (= 18). Thus, the maximum value of b that x_1 can assume is, b = Min (4, 18/3) = 4. Therefore,

$$f_1(4, 6, 18) = \underset{0 \le x_1 \le 4}{\text{Max}} \{3x_1\}$$

$$= 3 \text{ Min} \left\{4, 6 - \frac{2}{4}.x_2\right\}$$

and $x_1^* = \text{Min} \left\{4, 6 - \frac{2}{3}.x_2\right\}$

The recursive relation for optimization of this two-stage problem is

$$f_2(b_1, b_2, b_3) = \underset{0 \le x_2 \le b}{\text{Max}} \{5x_2 + f_1^*(b_1, b_2 - x_2, b_3 - 2x_2\}$$

where the maximization of x_2 satisfying the conditions of $x_2 \le b_2$ (= 6) and $2x_3$ (= 18) is the minimum of b = min (6, 9) = 6. Therefore, the recurrence relationship can be expressed as:

$$f_2(4, 6, 18) = \underset{0 \le x_2 \le 6}{\text{Max}} \{5x_2 = f_1^*(4, 6 - x_2, 18 - 2x_2)\}$$

$$= \underset{0 \le x_2 \le 6}{\text{Max}} \left\{5x_2 + 3\text{Min}\left(4, 6\frac{2}{3}x_2\right)\right\}$$

Since $\text{Min}\left(4, 6 - \frac{2}{3}.x_2\right)$

$$= \begin{cases} 4; & 0 \le x_2 \le 3 \\ 6 - \frac{2}{3}x_2; & 3 \le x_2 \le 6 \end{cases} = \begin{cases} 4; & 0 \le x_2 \le 3 \\ 6 - \frac{2}{3}x_2; & 3 \le x_2 \le 6 \end{cases}$$

we get $\text{Max} \left\{5x_2 + 3\text{Min}\left(4, 6 - \frac{2}{3}x_2\right)\right\} = \begin{cases} 5x_2 + 12; & 0 \le x_2 \le 3 \\ 18 + 3x_2; & 3 \le x_2 \le 6 \end{cases}$

Now the maximum value of $5x_2 + 12 = 27$ at $x_2 = 3$ and maximum value of $18 + 3x_2 = 36$ at $x_2 = 6$. Therefore, the optimal value of f_2^* = (4, 6, 18) = 36 is obtained at $x_2 = 6$. Since

$$x_1^* = \text{Min}\left\{4, 6 - \frac{2}{3}x_2\right\} = \text{Min}\left\{4, 6 - \frac{2}{3}\times 6\right\} = 2$$

The optimum solution to the given LP problem is; $x_1 = 2$, $x_2 = 6$ and Max Z = 36.

Example 13:

Solve the following LP problem by dynamic programming approach

Maximize $Z = 8x_1 + 7x_2$

subject to the constraints

$$2x_1 + x_2 \le 8$$

$$5x_2 + 2x_2 \le 15$$

and $x_1, x_2 \ge 0$

Solution:

This LP problem can be considered as a two-stage, two-state problem because there are two decision variables and two constraints. Starting with the second stage backward, the procedure is as follows:

The optimal value of $f_2(b_1, b_2)$ at the second stage is given by

$$f_2(b_1, b_2) = \underset{0 \le x_2 \le b}{\text{Max}} \{7x_2\}$$

where $b_1 = 8$, and $b_2 = 15$. The feasible value of x_2 is a non-negative value which satisfies all the given constraints $x_2 \le b_1$ (= 8) and $2x_2 \le b_2$ (= 15). Thus the maximum value of b that x_2 can assume is: b = min (8, 15/2) = 15/2. Therefore

$$f_2(b_1, b_2) = \underset{0 \le x_2 \le b}{\text{Max}} \{7x_2\}$$

$$= 7\ \text{Min}\ \{8 - 2x_1, (15/2) - (5/2)\ x_1\}$$

and $x_2^* = \text{Min}\ \{8 - x_4, 7.5 - 2.5\ x_1\}$

Proceeding backwards to stage 1 (j = 1), the recursive relation for optimization can be expressed as:

$$f_1(b_1, b_2) = \underset{0 \le x_1 \le b}{\text{Max}} \{8x_1 + f_2^*(b_1 - 2x_1, b_2 - (5/2)x_1)\}$$

$$= \underset{0 \le x_1 \le 3}{\text{Max}} \{8x_1 + 7\ \text{Min}\ (8 - 2x_1; (15/2) - (5/2)x_1)\}$$

where maximization of variable x_1 satisfying the conditions: $2x_1 \leq b$ (= 8) and $5x_1 \leq b_2$ (= 15) is the minimum of b = min (8/2, 15/5) = 3. Since the minimum (*i.e.*, zero) of $(8 - 2x_1, 15/2 - 5x_1/2)$ is obtained at x_1 = 3 for $0 \leq x_1 \leq 3$, we get

$$f_1^* (b_1, b_2) = \text{Max } [8x_1 + 7 \text{ Min } \{8 - 2x_1; (15/2) - (5x_1/2)\}]$$
$$= \text{Miax } \{8 \times 3 + 7 \times 0\} = 24, \text{ at } x_1 = 3$$

$$\text{and } x_2^* = \text{Min } \{8 = 2x_1, 7.5 - 2.5x_1\}$$
$$= \text{Min } \{8 - 6, 7.5 - 2.5 (3)\} = 0$$

Hence, the optimum solution to the given LP problem is: $x_1 = 3$, $x_2 = 0$ and Max Z = 94.

Example 14:

If $x_1 = 1$, $x_2 = 0$, $x_3 = 1$ be a feasible solution of the L.P.P.

$$\text{Min. } Z = 2x_1 + 3x_2 + 4x_3$$

subject to $x_1 + x_2 + x_3 = 2$

$$x_1 - x_2 + x_3 = 0,\ x_1, x_2, x_3 \geq 0$$

then show that the given feasible solution is not basic.

Solution:

The given system of constraint equations can be written in matrix form A x = b

$$\Rightarrow \quad \begin{bmatrix} 1 & 1 & 1 \\ 1 & -1 & 1 \end{bmatrix} \begin{bmatrix} x_1 \\ x_2 \\ x_3 \end{bmatrix} = \begin{bmatrix} 2 \\ 0 \end{bmatrix}.$$

$$\therefore \quad \alpha_1 = \begin{bmatrix} 1 \\ 1 \end{bmatrix}, \alpha_2 = \begin{bmatrix} 1 \\ -1 \end{bmatrix}, \alpha_3 = \begin{bmatrix} 1 \\ 1 \end{bmatrix}, b = \begin{bmatrix} 2 \\ 0 \end{bmatrix}.$$

The given feasible solution $x_1 = 1$, $x_2 = 0$, $x_3 = 1$ will not be basic if the vectors associated to non-zero variables are not linearly independent *i.e.*, if α_1 and α_3 are linearly dependent.

If we choose $\lambda_1 = -1$ and $\lambda_2 = 1$, we have

$$\lambda_1 \alpha_1 + \lambda_2 \alpha_2 = -1 \begin{bmatrix} 1 \\ 1 \end{bmatrix} + 1 \begin{bmatrix} 1 \\ 1 \end{bmatrix} = \begin{bmatrix} -1+1 \\ -1+1 \end{bmatrix} = \begin{bmatrix} 0 \\ 0 \end{bmatrix} = 0.$$

$\therefore$ the vectors α_1 and α_3 are linearly dependent.

Hence, the given feasible solution is not basic.

Example 15:

Maximize $Z = 2x_1 + 5x_2 + 7x_3$

subject to

$$3x_1 + 2x_2 + 4x_3 \leq 100$$

$$x_1 + 4x_2 + 2x_3 \leq 100$$

$$x_1 + x_2 + 3x_3 \leq 100,$$

$$x_1, x_2, x_3 \geq 0.$$

Solution:

The given problem is of maximization and all b_i's are positive.

Introducing slack variables x_4, x_5, x_6 to convert constraint inequalities into equations, the given problem becomes

$$Z = 2x_1 + 5x_2 + 7x_3 + 0x_4 + 0x_5 + 0x_6$$

	c_j	2	5	7	0	0	0	Min. ratio	
B	c_B	x_B	Y_1	Y_2	Y_3	Y_4	Y_5	Y_6	x_B/Y_3
Y_4	0	100	3	2	(4)	1	0	0	25
									←
Y_5	0	100	1	4	2	0	1	0	50
Y_6	0	100	1	1	3	0	0	1	100/3
$Z=c_B$	$x_B=0$	Δ_j	2	5	7	0	0	0	
					↑	↓			
Y_3	0	25	3/4	1/2	1	1/4	0	0	50
Y_5	0	50	1/2	(3)	0	–1/2	1	0	50/3
									←
Y_6	0	25	–5/4	–1/2	0	–3/4	0	1	Neg.
$Z=c_B$	$x_B=175$	Δ_j	–13/4	3/2	0	–7/4	0	0	
				↑			↓		
Y_5	7	50/3	5/6	0	1	1/3	–1/6	0	
Y_2	5	50/3	–1/6	1	0	–1/6	1/3	0	
Y_6	0	100/3	–4/3	0	0	–5/6	1/6	1	
$Z=c_Bx_B$ = 200		Δ_j	–3	0	0	–3/2	–1/2	0	

subject to $3x_1 + 2x_2 + 4x_3 + x_4 = 100$

$x_1 + 4x_2 + 2x_3 - x_5 = 100,\ x_1 + x_2 + 3x_3 + x_6 = 100$

The starting B.F.S. is

$x_1 = 0,\ x_2 = 0,\ x_3 = 0,\ x_4 = 100,\ x_5 = 100,\ x_6 = 100.$

The solution to the problem using simplex method is given below:

In the last all Δ_j's are zero or negative therefore the solution is optimal.

Hence the optimal solution is

$x_1 = 0,\ x_2 = 50/3,\ x_3 = 50/3$ and Max. $Z = 200$.

Example 16:

Max. $Z = x_1 + 2x_2 + 3x_3 - x_4$

subject to $x_1 + 2x_2 + 3x_3 = 15,\ 2x_1 + x_2 + 5x_3 = 20$

$x_1 + 2x_2 + x_3 + x_4 = 10,\ x_1, x_2, x_3, x_4 \geq 0.$

Solution:

The given problem is of maximization and all b_i's are positive. Also the constraints are equations. Examining the constraints we observe that in order to obtain a unit matrix of order 3 we need two more unit vectors as one unit vector is formed by the coefficients of x_4.

Therefore, introducing two artificial variables A_1 and A_2 in the first two constraints and assigning large negative penalty –M to the artificial variables, the given problem becomes

Max. $Z = x_1 + 2x_2 + 3x_3 - x_4 - MA_1 - MA_2$

subject to $x_1 + 2x_2 + 3x_3 + A_1 = 15$

$2x_1 + x_2 + 5x_3 + A_2 = 20,\ x_1 + 2x_2 + x_3 + x_4 = 10.$

The starting basic feasible solution is

$x_1 = 0,\ x_2 = 0,\ A_1 = 15,\ A_2 = 20,\ x_4 = 10.$

The solution to the problem using simplex method is given below:

Since all Δ_j's are negative or zero therefore the solution is optimal.

Hence the optimal solution is

$x_1 = 5/2 = x_2 = x_3,$

$x_4 = 0$ and Max. $Z = 15$.

B	c_j / c_B	1 / x_B	2 / Y_1	3 / Y_2	–1 / Y_3	–M / Y_4	–M / A_1	A_2	Min. ratio / x_B/Y^3
A^1	–M	15	1	2	3	0	1	0	5
A_2	–M	20	2	1	(5)	0	0	1	4 (Mini) ←
Y^4	–1	10	1	2	1	1	0	0	10
$Z=c_B x_B$ = –35M – 10		Δ_j	3M+2	3M+4	8M+4 ↓	0	0	0 ↓	x_B/Y_2
A_1 (Mini)	–M	3	–1/5	(7/5)	0	0	1		15/7 ←
Y_3	3	–4	2/5	1/5	1	0	0		20
Y_4	–1	6	3/5	9/5	0	1	0		10/3
$Z=c_B$ = –3M+6	x_B	Δ_j	$\frac{(-M+2)}{5}$	$\frac{7M+16}{5}$ ↑	0	0	0 ↓		x_B/Y_1
Y_2	2	15/7	–1/7	1	0	0			Neg.
Y_3	3	25/7	3/7	0	1	0			25/3
Y_4	–1	15/7	(6/7)	0	0	1			5/2 (Mini) ←
$Z=c_B$ = 90/7	x_B	Δ_j	6/7 ↑	0	0	0 ↓			
Y_2	2	5/2	0	1	0	1/6			
Y_3	3	5/2	0	0	1	–1/2			
Y_1	1	5/2	1	0	0	7/6			
$Z=c_B$ = 15	x_B	Δ_j	0	0	0	–1			

Example 17(a):

If $x_1 = 2,\ x_2 = 3,\ x_3 = 1$ *be a feasible solution of the linear programming problem*

$$\text{Max.} \quad Z = x_1 + 2x_2 + 4x_3$$

$$\text{s.t.} \quad 2x_1 + x_2 + 4x_3 = 11$$

$$3x_1 + x_2 + 5x_3 = 14$$

$$x_1, x_2, x_3 \geq 0.$$

then find a basic feasible solution from the given feasible solution.

Solution:

The given L.P.P. can be expressed as

Max. $Z = x_1 + 2x_2 + 4x_3$,

s.t. $A x = b$

$$\Rightarrow \quad \begin{bmatrix} 2 & 1 & 4 \\ 3 & 1 & 5 \end{bmatrix} \begin{bmatrix} x_1 \\ x_2 \\ x_3 \end{bmatrix} = \begin{bmatrix} 11 \\ 14 \end{bmatrix}$$

$$\Rightarrow \quad \alpha_1 x_1 + \alpha_2 x_2 + \alpha 3\, x_3 = b, \qquad \text{...(1)}$$

where $\alpha_1 = \begin{bmatrix} 2 \\ 3 \end{bmatrix}$, $\alpha_2 = \begin{bmatrix} 1 \\ 1 \end{bmatrix}$, $\alpha_3 = \begin{bmatrix} 4 \\ 5 \end{bmatrix}$, $b = \begin{bmatrix} 11 \\ 14 \end{bmatrix}$.

Since $x1 = 2$, $x_2 = 3$, $x_3 = 1$ is a feasible solution to the given L.P.P. therefore from (1)

$$2\alpha_1 + 3\alpha_2 + \alpha_3 = b.$$

Now the vectors α_1, α_2, α_3 associated with non-zero variables x_1, x_2, x_3 will be linearly dependent if one of these vectors can be expressed as the linear combination of the remaining two vectors.

Let $\lambda_1 \alpha_1 + \lambda_2 \alpha_2 = \alpha_3$ ...(2)

$$\Rightarrow \quad \lambda_1 \begin{bmatrix} 2 \\ 3 \end{bmatrix} + \lambda_2 \begin{bmatrix} 1 \\ 1 \end{bmatrix} = \begin{bmatrix} 4 \\ 5 \end{bmatrix}$$

$$\Rightarrow \quad \begin{bmatrix} 2\lambda_1 + \lambda_2 \\ 3\lambda_1 + \lambda_2 \end{bmatrix} = \begin{bmatrix} 4 \\ 5 \end{bmatrix}$$

$$\Rightarrow \quad 2\lambda_1 + \lambda_2 = 4,\ 3\lambda_1 + \lambda_2 = 5.$$

Solving these, we get

$$\lambda_1 = 1,\ \lambda_2 = 2.$$

Substituting the values of λ_1 and λ_2 in (2), we get

$$\alpha_1 + 2\alpha_2 = \alpha_3$$

$$\Rightarrow \quad \alpha_1 + 2\alpha_2 - \alpha_3 = 0$$

or $$\sum_{i=1}^{3} \lambda_i \alpha_i = 0$$

which gives $\lambda_1 = 1$,

$$\lambda_2 = 2, \lambda_3 = -1.$$

Now, we shall determine which of the three variables x_1, x_2, x_3 should be zero.

Using $v = \max_{1 \le i \le 3}\left(\frac{\lambda_i}{x_i}\right)$,

$$\Rightarrow \qquad v = \max\left(\frac{\lambda_i}{x_1}, \frac{\lambda_2}{x_2}, \frac{\lambda_3}{x_3}\right),$$

$$\Rightarrow \qquad v = \max\left(\frac{1}{2}, \frac{2}{3}, -\frac{1}{1}\right) = \frac{2}{3} = \frac{\lambda_2}{x_2}.$$

Since $x_i - \frac{\lambda_i}{v} \ge 0$, therefore

$$\left(x_i - \frac{\lambda_1}{v}, x_2 - \frac{\lambda_2}{v}, x_3 - \frac{\lambda_3}{v}\right)$$

is the reduced feasible solution. We have

$$x_1 - \frac{\lambda_1}{v} = 2 - \frac{1}{2} = \frac{1}{2},$$

$$x_2 - \frac{\lambda_2}{v} = 0,$$

$$\Rightarrow \qquad x_3 - \frac{\lambda_3}{v} = \frac{5}{2}.$$

$\therefore$ The new feasible solution is $x_1 = \frac{1}{2}$,

$$x_3 = 0, x_3 = \frac{5}{2}.$$

Now $\alpha_1 = \begin{bmatrix} 2 \\ 3 \end{bmatrix}$, and $\alpha_3 = \begin{bmatrix} 4 \\ 5 \end{bmatrix}$ are the vectors associated to non-zero variables x_1 and x_3.

Since α_1 and α_3 are linearly independent vectors therefore the new feasible solution

$$x_1 = \frac{1}{2}, x_2 = 0, x_3 = \frac{5}{2}$$

is the basic feasible solution.

Note: In order to find another basic feasible solution we proceed as follows:

Since α_1, α_2, α_3 are L.D. therefore one of them can be expressed as the linear combination of the other two.

Let $\lambda_2 \alpha_2 + \lambda_3 \alpha_3 = \alpha_1$. ...(3)

Now proceed as in Example 9.

We have $\lambda_1 = -1$, $\lambda_2 = -2$,

$$\lambda_3 = 1, \; v = \frac{1}{1} = \frac{\lambda_3}{x_3}.$$

B.F.S. is $x_1 = 3$, $x_2 = 5$, $x_3 = 0$.

Example 17(b):

Consider the set of equations

$$5x_1 - 4x_2 + 3x_3 + x_4 = 3$$
$$2x_1 + x_2 + 5x_3 - 3x_4 = 0$$
$$x_1 + 6x_2 - 4x_3 + 2x_4 = 15,$$
$$x_1, x_2, x_3, x_4 \geq 0.$$

If $x_1 = 1$, $x_2 = 2$, $x_3 = 1$, $x_4 = 3$ is a feasible solution then find a basic feasible solution.

Solution:

The given set of equations can be expressed in the matrix form

$$A\,x = b$$

$$\Rightarrow \begin{bmatrix} 5 & -4 & 3 & 1 \\ 2 & 1 & 5 & -3 \\ 1 & 6 & -4 & 2 \end{bmatrix} \begin{bmatrix} x_1 \\ x_2 \\ x_3 \\ x_4 \end{bmatrix} = \begin{bmatrix} 3 \\ 0 \\ 15 \end{bmatrix}$$

$$\Rightarrow \alpha_1 x_1 + \alpha_2 x_2 + \alpha_3 x_3 + \alpha_4 x_4 = b \qquad ...(1)$$

where $\alpha_1 = \begin{bmatrix} 5 \\ 2 \\ 1 \end{bmatrix}, \alpha_2 = \begin{bmatrix} -4 \\ 1 \\ 6 \end{bmatrix}, \alpha_3 = \begin{bmatrix} -3 \\ 5 \\ -4 \end{bmatrix}, \alpha_4 = \begin{bmatrix} 1 \\ -3 \\ 2 \end{bmatrix}, b = \begin{bmatrix} 3 \\ 0 \\ 15 \end{bmatrix}.$

Since $x_1 = 1$, $x_2 = 2$, $x_3 = 1$, $x_4 = 3$ is a feasible solution to the given set of equations, therefore from (1)

$$\alpha_1 + 2\alpha_2 + \alpha_3 + 3\alpha_4 = b. \qquad ...(2)$$

Now the vectors α_1, α_2, α_3, α_4 associated with non-zero variables x_1, x_2, x_3, x_4 will be linearly dependent if one of these vectors can be expressed as the linear combination of the remaining three vectors.

Let $\lambda_2 \alpha_2 + \lambda_3 \alpha_3 + \lambda_4 \alpha_4 = \alpha_1$ ...(3)

$$\Rightarrow \lambda_2 \begin{bmatrix} -4 \\ 1 \\ 6 \end{bmatrix} + \lambda_3 \begin{bmatrix} 3 \\ 5 \\ -4 \end{bmatrix} + \lambda_4 \begin{bmatrix} 1 \\ -3 \\ 2 \end{bmatrix} = \begin{bmatrix} 5 \\ 2 \\ 1 \end{bmatrix}$$

$$\Rightarrow \begin{bmatrix} -4\lambda_2 + 3\lambda_3 + \lambda_4 \\ \lambda_2 + 5\lambda_3 - 3\lambda_4 \\ 6\lambda_2 - 4\lambda_3 + 2\lambda_4 \end{bmatrix} = \begin{bmatrix} 5 \\ 2 \\ 1 \end{bmatrix}$$

$$\Rightarrow -4\lambda_2 + 3\lambda_3 + \lambda_4 = 5$$

$$\lambda_2 + 5\lambda_3 - 3\lambda_4 = 2$$

$$6\lambda_2 - 4\lambda_3 + 2\lambda_4 = 1.$$

Solving these, we get

$$\lambda_2 = \frac{22}{43}, \lambda_3 = \frac{139}{86}, \lambda_4 = \frac{189}{86}.$$

Substituting these values in (3), we get

$$\frac{22}{43}\alpha_2 + \frac{139}{86}\alpha_3 + \frac{189}{86}\alpha_4 = \alpha_1$$

$$\Rightarrow 86\alpha_1 - 44\alpha_2 - 139\alpha_3 - 189\alpha_4 = 0 \qquad ...(4)$$

$$\Rightarrow \sum_{i=1}^{4} \lambda_i \alpha_i = 0, \text{ which gives}$$

$$\lambda_1 = 86, \lambda_2 = -44, \lambda_3 = -139, \lambda_4 = -189.$$

Using $v = \max_{1 \le i \le 4} \left(\frac{\lambda_i}{x_i} \right)$, we have

$$v = \max \left(\frac{\lambda_1}{x_1}, \frac{\lambda_2}{x_2}, \frac{\lambda_3}{x_3}, \frac{\lambda_4}{x_4} \right)$$

$$= \max \left(\frac{86}{1}, -\frac{44}{2}, -\frac{139}{1}, \frac{-189}{3} \right) = \frac{86}{1} = \frac{\lambda_1}{x_1}.$$

$\therefore$ the variable x_1 should be zero

i.e., the vector a1 should be eliminated.

Eliminating α_1 between (2) and (4), we get

$$\frac{44\alpha_2 + 139\alpha_3 + 189\alpha_4}{86} + 2\alpha_2 + \alpha_3 + 3\alpha_4 = b$$

$$\Rightarrow \quad 0.\alpha_1 + \frac{216}{86}\alpha_2 + \frac{225}{86}\alpha_3 + \frac{447}{86}\alpha_4 = b$$

$$\therefore \quad x_1 = 0,\ x_2 = \frac{216}{86},\ x_3 = \frac{225}{86},\ x_4 = \frac{447}{86}$$

is the new feasible solution.

Since the vectors

$$\alpha_2 = \begin{bmatrix} -4 \\ 1 \\ 6 \end{bmatrix},\ \alpha_3 = \begin{bmatrix} 3 \\ 5 \\ -4 \end{bmatrix},\ \alpha_4 = \begin{bmatrix} 1 \\ -3 \\ 2 \end{bmatrix}$$

associated to non-zero variables are L.I. therefore the new feasible solution is the basic feasible solution.

Note: Another feasible solution is

$$x_1 = \frac{25}{139},\ x_2 = \frac{234}{139},\ x_3 = 0,\ x_4 = \frac{228}{139}.$$

Example 17(c):

Solve the following system of linear equations by using simplex method

$$x_1 - x_3 + 4x_4 = 3,\ 2x_1 - x_2 = 3$$

$$3x_1 - 2x_2 - x_4 = 1,\ x_1, x_2, x_3, x_4 \geq 0.$$

Solution:

Introducing the dummy objective function Z with costs zero to each given variable and cost (–1) to each artificial variable the given equations can be written as a linear programming problem in the following form

$$\text{Max. } Z = 0x_1 + 0x_2 + 0x_3 + 0x_4 - A_1 - A_2 - A_3$$

subject to

$$x_1 - x_3 + 4x_4 + A_1 = 3$$

$$2x_1 - x_2 + A_2 = 3$$

$$3x_1 - 2x_2 - x_4 + A_3 = 1,$$

$$x_1, x_2, x_3, x_4, A_1, A_2, A_3 \geq 0.$$

The starting B.F.S. is

$x_1 = 0,\ x_2 = 0,\ x_3 = 0,$

$x_4 = 0,\ A_1 = 3,\ A_2 = 3,\ A_3 = 1.$

The solution to the problem using simplex algorithm is

B	c_j / c_B	0 / x_B	0 / Y_1	0 / Y_2	0 / Y_3	−1 / Y_4	−1 / A_1	−1 / A_2	A_3	Min. ratio / x_B/Y_1
A_1	−1	3	1	0	−1	4	1	0	0	3
A_2	−1	3	2	−1	0	0	0	1	0	3/2
A_3	−1	1	(3)	−2	0	−1	0	0	1	1/3
										←
$Z = c_B x_B$	Δ_j		6	−3	−1	3	0	0	0	X_B/Y_4
= −7			↑						↓	
A_1	−1	8/3	0	2/3	−1	(13/3)	1	0		8/13
A_2	−1	7/3	0	1/3	0	2/3	0	1		7/2
Y_1	0	1/3	1	−2/3	0	−1/3	0	0		Neg.
$Z = c_B x_B$	Δ_j		0	1	−1	5	0	0		
= − 5						↑	↓			
Y_4	0	8/13	0	(2/13)	−3/13	1		0		4
										←
A_2	−1	25/13	0	3/13	2/13	0		1		25/3
Y_1	0	7/13	1	−8/13	−1/13	0		0		Neg.
−25/13		Δ_j	1	3/13	2/13	0		0		x_B/Y_3
			↑		↓					
Y_2	0	4	0	1	−3/2	13/2		0		Net.
A_2	−1	1	0	0	(1/2)	−3/2		1		2
										←
Y_1	0	3	1	0	−1	4		0		Neg.
Z= −1		Δ_j	0	0	1/2	−3/2		0		
					↑			↓		
Y_2	0	7	0	1	0	2				
Y_3	0	2	0	0	1	−3				
Y_1	0	5	1	0	0	1				
Z=0	Δ_j	0	0	0	0					

Since all $\Delta_j \leq 0$ therefore the problem has optimal solution given Since all Δ_j's are negative or zero therefore the solution is optimal.

Hence the optimal solution is

$x_1 = 5/2 = x_2 = x_3, x_4 = 0$ and Max. $Z = 15$. by

$x_1 = 5, x_2 = 7,$

$x_3 = 2, x_4 = 0.$

Example 17(d):

Apply simplex method to find the inverse of the matrix $\begin{bmatrix} 4 & 3 \\ 3 & 2 \end{bmatrix}$.

Solution:

Let $A = \begin{bmatrix} 4 & 3 \\ 3 & 2 \end{bmatrix}$ and

$b = \begin{bmatrix} 4 \\ 6 \end{bmatrix}$ be a dummy real vector.

Consider the system of equations

$A\,x = b$ given by $\begin{bmatrix} 4 & 3 \\ 3 & 2 \end{bmatrix}\begin{bmatrix} x_1 \\ x_2 \end{bmatrix} = \begin{bmatrix} 4 \\ 6 \end{bmatrix}$, $x_1, x_2 \geq 0$.

Introducing the artificial variables x_3, x_4 and the dummy objective function Z, the linear programming problem becomes

		c_j	0	0	–1	–1	**Min. ratio**
B	c_B	x_B	Y_1	Y_2	A_1	A_2	x_B/Y_1
A_1	–1	4	(4)	3	1	0	4/4 = 1 ←
A_2	–1	6	3	2	0	1	6/3 = 2
$Z=c_B$	$x_B = -10$	Δ_j	7 ↑	5	0 ↓	0	
Y_1	0	1	1	3/4	1/4	0	
A_2	–1	3	0	(1/4)	–3/4	1	
$Z=c_B$	$x_B = -3$	Δ_j	0	–1/4 ↑	–7/4	0	

Max. $Z = 0x_1 + 0x_2 - x_3 - x_4$

s.t. $\quad 4x_1 + 3x_2 + x_3 = 4$

$$3x_1 + 2x_2 + x_4 = 6,$$

$$x_1, x_2, x_3, x_4 \geq 0.$$

The starting B.F.S. is $x_3 = 4$, $x_4 = 6$.

Since all Δ_j's are negative or zero therefore the solution is optimal. But the matrix A is still not converted into the unit matrix. For this introduce Y_2 into the basis and drop A_2. The new simplex table becomes

B	c_B	c_j / x_B	0 / Y_1	0 / Y_2	–1 / A_1	–1 / A_2
Y_1	0	10	1	0	–2	3
Y_2	0	–12	0	1	3	–4
$Z=c_B$	$x_B=0$	Δ_j	0	0	–1	–1

Since the columns of the given matrix A have become the columns of the unit matrix I therefore

$$A^{-1} = \begin{bmatrix} -2 & 3 \\ 3 & -4 \end{bmatrix}.$$

Example 18:

Find the inverse of the matrix

$$B_\alpha = (\beta_1, \beta_2, \alpha)$$

where $$B = (\beta_1, \beta_2, \beta_3),$$

$$\alpha = \begin{bmatrix} 2 \\ 1 \\ 3 \end{bmatrix} \text{ and } B^{-1} = \begin{bmatrix} 7 & -3 & -3 \\ -1 & 1 & 0 \\ -1 & 0 & 1 \end{bmatrix}.$$

Solution:

We know that $Y = (y_1, y_2, y_3) = B^{-1}\alpha$

$$\therefore \quad Y = \begin{bmatrix} 7 & -3 & -3 \\ -1 & 1 & 0 \\ -1 & 0 & 1 \end{bmatrix}\begin{bmatrix} 2 \\ 1 \\ 3 \end{bmatrix} = \begin{bmatrix} 2 \\ -1 \\ 1 \end{bmatrix}$$

or $\quad y_1 = 2, y_2 = -1, y_3 = 1.$

$$\text{Now } A_3 = \left(-\frac{y_1}{y_3}, -\frac{y_2}{y_3}, \frac{1}{y_3}\right) = (-2, 1, 1)$$

$$\therefore \quad E = (e_1, e_2, A_3) = \begin{bmatrix} 1 & 0 & -2 \\ 0 & 1 & 1 \\ 0 & 0 & 1 \end{bmatrix}$$

Now $$B_\alpha^{-1} = EB^{-1} = \begin{bmatrix} 1 & 0 & -2 \\ 0 & 1 & 1 \\ 0 & 0 & 1 \end{bmatrix} \begin{bmatrix} 7 & -3 & -3 \\ -1 & 1 & 0 \\ -1 & 0 & 1 \end{bmatrix}$$

$$= \begin{bmatrix} 9 & -3 & -5 \\ -2 & 1 & 1 \\ -1 & 0 & 1 \end{bmatrix}.$$

Example 19:

Max. $Z = 2x_1 + 3x_2$

subject to $-x_1 + 2x_2 \leq 4$

$x_1 + x_2 \leq 6$

$x_1 + 3x_2 \leq 9$, x_1, x_2 *are unrestricted.*

Solution:

The given problem is of maximization and all b_i's are positive. We know that the simplex method is applicable if the variables are non-negative. In the given problem x_1 and x_2 are unrestricted (may be + , – or zero). *In such cases we replace all these variables by the difference of two non-negative variables and solve the problem in usual manner.*

$\therefore$ Making the transformation

$$x_1 = x_1' - x_1'' \text{ and } x_2 = x_2' - x_2'$$

such that $x_1', x_2'', x_2', x_2'' \geq 0$.

The given problem becomes

$$\text{Max. } Z = 2(x_1' - x_1'') + 3(x_2' - x_2'')$$

s.t. $-(x_1' - x_1'') + 2(x_2' - x_2'') \leq 4$

$(x_1' - x_1'') + (x_2' - x_2'') \leq 6$

$(x_1' - x_1'') + 3(x_2' - x_2'') \leq 9$

$x_1', x_2', x_1'', x_2'' \geq 0.$

To change constraint inequalities into equations introducing slack variables x_3, x_4 and x_5, the given L.P.P. becomes

$$\text{Max. } Z = 2x_1' - 2x_1'' + 3x_2' - 3x_2'' + 0.x_3 + 0.x_4 + 0.x_5$$

$$-x_1' + x_1'' + 2x_2' - 2x_2'' + x_3 = 4$$

$$x_1' - x_1'' + x_2' - x_2'' + x_4 = 6$$
$$x_1' - x_1'' + 3x_2' - 3x_2'' + x_5 = 9.$$

The starting B.F.S. is $x_3 = 4$, $x_4 = 6$, $x_5 = 9$.

The solution to the problem using simplex algorithm is given below:

B	c_j c_B	2 x_B	–2 Y_1'	3 Y_1''	–3 Y_2'	0 Y_2''	0 Y_3	0 Y_4	Min. ratio Y_5	x_B/Y_2'
Y_3	0	4	–1	1	(2)	–2	1	0	0	2 ←
Y^4	0	6	1	–1	1	–1	0	1	0	6
Y^5	0	9	1	–1	3	–3	0	0	1	3
$Z=c^B$ =0	x^B	Δ_j	2	–2	3 ↑	–3	0 ↓	0	0	
Y_2'	3	2	–1/2	1/2	1	–1	1/2	0	0	Neg.
y_4	0	4	3/2	–3/2	0	0	–1/2	1	0	8/3
Y_5	0	3	(5/2)	–5/2	0	0	–3/2	0	1	6/5 ←
$Z=c_B$ =6	x_B	Δ_j	7/2 ↑	–7/2	0	0	–3/2	0	0 ↓	xB/Y3
y_2'	3	13/5	0	0	1	–1	1/5	0	1/5	13
Y_4	0	11/5	0	0	0	0	(2/5)	1	–3/5	11/2 ←
Y_1'	2	6/5	1	–1	0	0	–3/5	0	2/5	Neg.
$Z=c_B$ =51/5	x_B	Δ_j	0	0	0	0	3/5 ↑	0 ↓	–7/5	
Y_2'	3	3/2	0	0	1	–1	0	–1/2	1/2	
Y_3	0	11/2	0	0	0	0	1	5/2	–3/2	
Y_1'	2	9/2	1	–1	0	0	0	3/2	–1/2	
$Z=c_B$ =27/2	x_B	Δ_j	0	0	0	0	–3/2	–1/2		

Since all Δ_j's are negative or zero therefore the solution is optimal.

∴ The optimal solution is

$x_1' = 9/2,\ x_1'' = 0,\ x_2' = 3/2,\ x_2'' = 0,$

Max. $Z = 27/2$

$\Rightarrow\ x_1 = x_1' - x_1'' = 9/2,$

$x_2 = x_2' - x_2'' = 3/2,$

Max. $Z = 27/2.$

Example 20:

Apply Big-M method to solve the problem

$$Max\ Z = -x_1 - x_2$$

subject to

$$3x_1 + 2x_2 \geq 30$$

$$-2x_1 + 3x_2 \leq -30$$

$$x_1 + x_2 \leq 5,$$

$$x_1, x_2 \geq 0.$$

Solution:

The given problem is of maximization.

Since bi in second constraint is negative therefore multiplying both sides by – 1, we et

$$2x_1 - 3x_2 \geq 30.$$

To convert inequalities of constraints into equations, introducing slack, surplus and artificial variables wherever necessary the given problem becomes

$$\text{Max. } Z = -x_1 - x_2 + 0x_3 + 0x_4 + 0x_5 - MA_1 - MA_2$$

subject to

$$3x_1 + 2x_2 - x_3 + A_1 = 30$$

$$2x_1 - 3x_2 - x_4 + A_2 = 30$$

$$x_1 + x_2 + x_5 = 5$$

where $x_1, x_2, x_3, x_4, x_5,$

$A_1, A_2 \geq 0.$

The initial B.F.S. is $x_1 = 0,\ x_2 = 0,\ x_3 = 0,$

$$x_4 = 0, x_5 = 5, A_1 = 30, A_2 = 30.$$

The solution to the problem using simplex method is given below:

	c_j	–1	–1	0	0	0	–M	–MMin. ratio		
B	c_B	x_B	Y_1	Y_2	Y_3	Y_4	Y_5	A_1	A_2	x_B/Y_1
A_1	–M	30	3	2	–1	0	0	1	0	10
A_2	–M	30	2	–3	0	–1	0	0	1	15
Y_5	0	5	(1)	1	0	0	1	0	0	5 ←
$Z=c_B\ x_B = -60M$		Δ_j	5M–1 ↑	–M	–M	–M	0 ↓	0	0	
A_1	–M	15	0	–1	–1	0	–3	1	0	
A_2	–M	20	0	–5	0	–1	–2	0	1	
Y_1	–1	5	1	1	0	0	1	0	0	
Z=–35 M–5		Δ_j	0	–6M	–M	–M	–5M	0	0	

Since all $\Delta_j \leq 0$ therefore the solution obtained is optimal. But the artificial variables A_1, A_2 appear in the basis at a positive level which implies that *the given L.P.P. has no feasible solution.*

Note: Sometimes the constraints may be inconsistent so that there is no feasible solution to the problem. Such a situation is called ***infeasibility***. *In the case of infeasibility one or more artificial variables appear in the final simplex table.*

Two Phase Method. As an alternative to the Big-M method, there is another method for dealing with linear programming problems involving artificial variables. This is called the two-dealing with linear and as its name suggests it separates the solution procedure into two phases. In phase-I, all the he artificial variables are eliminated from the basis. In phase II, we use the solution from phase I as the initial basic feasible solution and use the simplex method to determine the optimal solution.

Example 21:

Max. $Z = 10x_1 + 20x_2$

subject to $2x_1 + 4x_2$

$x_1 + 5x_2 \geq 15,$

$x_1, x_2 \geq 0.$

Solution:

The given problem is of maximization and all b_i's are positive.

Introducing surplus variables x_3, x_4 and artificial variables A_1, A_2, the given problem becomes

$$\text{Max. } Z = 10x_1 + 20x_2 + 0x_3 + 0x_4 - MA_1 - MA_2$$

subject to $2x_1 + 4x_2 - x_3 + A_1 = 16$

$x_1 + 5x_2 - x_4 + A_2 = 15,$

$x_1, x_2, x_3, x_4, A_1, A_2, \geq 0.$

The starting B.F.S. is

$x_1 = 0, x_2 = 0, x_3 = 0, x_4 = 0, A_1 = 16, A_2 = 15.$

The final simplex table in respect of this problem is

B	c_B	c_j / x_B	10 / Y_1	20 / Y_2	0 / Y_3	0 / Y_4	–M / A_1	–M / A_2	Min. ratio
Y_1	10	15	1	5	0	–1	0	1	
Y_3	0	14	0	6	1	–2	–1	2	
Z=150	Δ_j	0	–30	0	10 ↑	–M	–M – 10		

From the table we observe that Y_4 is the incoming vector but we can not find outgoing vector because all min. ratios are negative. Therefore the solution is unbounded.

Note: If in a situation there is at least one Δ_j greater than zero but there is no non-negative ratios or min. ratios $\rightarrow \infty$ then we have unbounded solution.

L.P.P. having more than one optimum solution.

Example 22(a):

Max. $Z = 6x_1 + 4x_2$

subject to $2x_1 + 3x_2 \leq 30$

$3x_1 + 2x_2 \leq 30$

$x_1 + x_2 \geq 3,\ x_1, x_2, \geq 0.$

Solution:

To convert inequalities into equations, introducing slack, surplus and artificial variables, the problem becomes

$$\text{Max. } Z = 6x_1 + 4x_2 + 0x_3 + 0x_4 + 0x_5 - MA_1$$

subject to $2x_1 + 3x_2 + x_3 = 30$

$3x_1 + 2x_2 + x_4 = 24$

$x_1 + x_2 - x_5 + A_1 = 3,$

$x_1, x_2, x_3, x_4, x_5, A_1 \geq 0.$

The starting B.F.S. is $x_1 = 0, x_2 = 0, x_3 = 30, x_4 = 24, x_5 = 0, A_1=3.$

The solution to the problem using simplex method is given below:

		c_j	6	4	0	0	0	–M	Min. ratio
B	c_B	x_B	Y_1	Y_2	Y_3	Y_4	Y_5	A_1	x_B/Y_1
Y_3	0	30	2	3	1	0	0	0	15
Y_4	0	24	3	2	0	1	0	0	8
A_1	–M	3	(1)	1	0	0	–1	1	3 ←
$Z=c_B\ x_B$ = –3M		Δ_j	6+M □↑	4+M	0	0	–M	0 ↓	x_B/Y_5
Y_3	0	24	0	1	1	0	2		12
Y_4	0	15	0	–1	0	1	(3)		5 ←
Y_1	6	3	1	1	0	0	–1		–
Z=18		Δ_j	0	–2	0	0 □↓	6 ↑		
Y_3	0	14	0	(5/3)	1	–2/3	0		
Y_5	0	5	0	–1/3	0	1/3	1		
Y_1	6	8	1	2/3	0	1/3	0		
Z=48		Δ_j	0	0	0	–2	0		

Since all $\Delta_j \leq 0$ therefore the solution is optimal. The optimal solution is $x_1 = 8$, $x_2 = 0$, max $Z = 48$.

Again $\Delta_2 = 0$ and Y_2 is not in the basis B therefore an alternative solution also exists. Thus the problem does not have unique solution.

Taking Y_2 as incoming and Y_3 as outgoing vector we obtain the following simplex table:

		c_j	6	4	0	0	0	Min. ratio
B	c_B	x_B	Y_1	Y_2	Y_3	Y_4	Y_5	
Y_2	4	42/5	0	1	3/5	–2/5	0	
Y_5	0	39/5	0	0	1/5	1/5	1	
Y_1	6	12/5	1	0	–2/5	3/5	0	
$Z = c_B\ x_B \Delta_j$ =48			0	0	0	–2	0	

All $\Delta_j \leq 0$ therefore this solution is also optimal having the same maximum value of Z.

Second optimal solution is

$x_1 = 12/5$, $x_2 = 42/5$ and

max $Z = 48$.

We know that the convex combination of B.F.S. is also an optimal solution.

Thus if we obtain two alternative optimum solution then we can obtain any number of optimum solutions.

For any arbitrary value of λ the following table gives different optimum solution:

Variables	I Sol.	II Sol.	Gen. Sol.
x_1	8	12/5	$8\lambda + (12/5)(1 - \lambda)$
x_2	0	42/5	$0\lambda + (42/5)(1 - \lambda)$
x_3	14	0	$14\lambda + 0(1 - \lambda)$
x_4	0	0	$0\lambda + (1 - \lambda)$
x_5	5	39/5	$5\lambda + (39/5)(1 - \lambda)$
A_1	0	0	$0\lambda + 0(1 - \lambda)$

Taking a particular value of λ, $\lambda = \frac{1}{2}$ (say), *the third optimal solution is*

$x_1 = 26/5,\ x_2 = 21/5$ and Max. $Z = 48$.

Example 22(b):

Max. $Z = 6x_1 - 2x_2$

subject to

$2x_1 - x_2 \leq 2$

$x_1 \leq 4$

$x_1,\ x_2 \geq 0.$

Solution:

The given problem is of maximization and all b_i's are positive.

To convert inequalities of constraints into equations introducing slack variables x_3 and x_4, given problem becomes

Max. $Z = 6x_1 - 2x_2 + 0x_3 + 0x_4$

subject to

$2x_1 - x_2 + x_3 = 2$

$x_1 + x_4 = 4.$

B	c_B	c_j x_B	**6** Y_1	**–2** Y_2	**0** Y_3	**0** Y_4	**Min. ratio** x_B/Y_1
Y_3	0	2	(2)	–1	1	0	1 ←
Y_4	0	4	1	0	0	1	4
$Z=c_B$	$x_B=0$	Δ_j	6 ↑	–2	0 ↓	0	x_B/Y_2
Y_1	6	1	1	–1/2	1/2	0	Neg.
Y_4	0	3	0	(1/2)	–1/2	1	6 ←
Z=6	Δ_j	0	1 ↑	–3	0 ↓		
Y_1	6	4	1	0	0	1	
Y_2	–2	6	0	1	–1	2	
Z=12	Δ_j	0	0	–2	–2		

The starting B.F.S. is

$x_1 = 0, x_2 = 0,$

$x_3 = 2, x_4 = 4.$

The solution to the problem using simplex method is given below:

Since all Δ_j's are negative or zero therefore the solution obtained is optimal.

The optimal solution is

$x_1 = 4, x_2 = 6$ and

max $Z = 12$.

From the first simplex table we observe that the elements of column of Y_2 are either negative or zero which indicates that the feasible region is not bounded.

Hence a linear programming problem having unbounded feasible solution may have a bounded optimal solution.

EXERCISES

1. Show that if a linear programming problem has a feasible solution, it also has a basic feasible solution.
2. State and prove the fundamental theorem of linear programming.
3. Food A contains 20 units of vitamin X and 40 units of vitamin Y per gram. Food B contains 30 units each of vitamin X and Y. The daily minimum human requirements of vitamins X and Y are 900 units and 1200 units respectively. How many grams of each type of food should be consumed so as to minimize the cost if food A costs 60 praise per gram and food B costs 80 praise per gram.
4. A finished product must weigh exactly 150 gms. The two raw materials used in manufacturing the product are : A with cost of Rs. 2 per unit and B with a cost of Rs. 8 per unit. At least 14 units of B and not moire than 20 units of A must be used. Each unit of A and B weighs 5 and 10 grams respectively.
5. Use dynamic programming to find the value of

   Max $Z = y_1 \cdot y_2 \cdot y_3$

   subject to the constraint

$y_1 + y_2 + y_3 = 5$

and $y_1, y_2, y_3 \geq 5$

6. Show how the functional equation technique of dynamic programming can be used to determine the shortest route when it is constrained to pass through a set of specified nodes which is a definite subset of the set of nodes of a given network.
7. State Bellman' principle of optimality and apply it to solve the following problem

   Max $Z = x_1 . x_2 ... x_n$

   subject to the constraint

   $x_1 + x_2 + ..., + x_n = c$

   and $x_1, x_2, + ..., + x_n \geq 0$
8. A government space project is conducting research on a certain engineering problem that must be solved before a man can fly to moon safely. These research teams are currently trying three different approaches for solving the problems. An estimate has been made that, under present circumstances, the probability that the respective teams; say A, B and C will not succeed are 0.40, 0.60 and 0.80, .respectively. Thus, the current probability that all three teams will fail is (0.40) (0 60) (0.80) = 0.192. Since the objective is to minimize this probability, the decision has been made to assign two or more top scientists among the three teams in order to lower the probability of failure as much as possible.

   The following table gives the estimated probability that the respective teams will fail when 0, 1 or 2 additional scientists are added to that team.

Number of new scienntists	Team A	Team B	Team C
0	0.40	0.60	0.80
1	0.20	0.40	0.50
2	0.15	0.20	0.30

   How should the additional scientists be allocated to the team?
9. A truck can carry a total of 10 tonnes of product. Three types of products are available for shipment. Their weights and values are tabulated. Assuming that at least one of each type must be

shipped, determine the loading which will maximize the total value.

Product Type	*Value (Rs)*	*Weight (Rs)*
A	20	1
B	50	2
C	60	3

10. Consider the problem of designing an electronic device consisting of three main components. The components are arranged in series so that the failure of one of them will result in the failure of the whole device. Therefore, it is decided that the reliability (probability of failure) of the device can be increased by installing parallel units on each component. Each component may be installed in at the most three parallel units. The total capital (in thousand Rs) available for the device is 10. The following data is available:

Number of Parallel Units, m_i	***Components***					
	1		***2***		***3***	
	r_1	c_1	r_2	c_2	r_3	c_3
1	0.50	2	0.70	3	0.60	1
2	0.70	4	0.80	5	0.80	2
3	0.90	5	0.90	6	0.90	3

where r_i, c_i (i = 1, 2, 3) is the reliability and the cost of the ith component, respectively. Determine the number of parallel units which will maximize the total reliability of the system without exceeding the given capital.

11. Use the principle of optimality to find the maximum value of

Max $Z = b_1x_1 + b_2x_2 + ... + b_nx_n$

subject to the constrain $= x_1 + x_2 + ... + x_n = c$

and $x_1, x_2, ..., x_n \geq 0$

12. A student has to take an examination in three courses x, y and z. He has three days available for study. He feels that it would be better to devote a whole day to study of the same course, so that he may study a course for one day, two days or three days or not at all. His estimates of grades he may get according to

days of study he puts in are as follows:

Study Days	*Course*		
	x	*y*	*z*
0	1	2	1
1	2	2	2
2	2	4	4
3	4	5	4

13. A chairman of a certain political party is making plans for his election to the Parliament. He has engaged the services of six volunteer workers and wishes to assign them four districts in such a way as to maximize their effectiveness. He feels that it would be inefficient to assign a worker to more than one district but he is also willing to assign no worker to any one of the districts judging by what the workers can accomplish in other districts.

    The following table gives the estimated increase in the number of votes in favour of the party's candidate if it were allocated various number of workers;

Number of Workers	*Districts* 1	2	3
0	0	0	0
1	25	20	33
2	42	38	43
3	55	54	47
4	63	65	50
5	69	73	52
6	74	80	53

How many of the workers should be assigned to each of the three districts in order to maximize the total number of votes in his favour?

14. We have a machine that deteriorates with age and so we need to formulate a replacement policy. We have to own such a machine during each of the next years. The operating cost c(i)

of a machine i years old at the beginning of the year, trade in value t(i) received when such a machine is traded for a new machine at the start of the year and s(i), the salvage value received for a machine that have just turned age i at the end of 5 years are given below:

i	:	0	1	2	3	4	5	6
c(i)	:	10	13	20	40	70	100	100
t(i)	:	–	32	21	11	5	0	0
s(i)	:	–	25	17	8	0	0	0

If a new machine costs Rs 50 and we have now a machine which is two years old, what is the optimum policy of replacement? Solve the problem by dynamic programming.

15. The following table provides the monthly demand for an item of a company during the winter season. Virtually all direct labour is casual and temporary so that any number of items up to the plant capacity of 400 items may be produced each month. The direct labour cost is Rs 40 per item.

Month	: November	December	January	February
Demand	: 100	200	300	400

The plant is completely shut down during a month when there is no production, and supervisory personnel are given an unpaid holiday to save a further Rs 2,000 in monthly payroll costs. Demands may be filled either from current production or from inventory. Each item held in inventory at a beginning of the month costs the company Rs 5. The production lot size may be 0, 100, 200, 300 or 400 items. Determine the monthly production levels in such a way that total cost is minimized,

16. A company is planning its manufacturing operations for the next five months. The following demands apply:

Month	: January	February	March	April	May
Demand	: 200	300	300	200	400

Each item held in inventory from one month to the next incurs a Rs 4 carrying charge. There are 200 unsold items remaining at the end of December. The company is to redesign its production for June, so that a no ending inventory is desired. A maximum of 400 units may be manufactured in any given month. The total production costs are:

Size of production	:	0	100	200	300	400
Cost (Rs)	:	0	1,000	1,300	1,450	1,525

Determine the minimum cost production plan.

17. A shoe store sells rubber shoes for a particular season which lasts from December 1 through February 29. The sales division has forecast the following demands for the next year.

Month	:	December	January	February
Demand	:	30	40	30

All shoes sold by the store are purchased from outside sources. The following information is known about this particular shoe.

(i) The unit purchasing cost is Rs 100 per pair; however, the supplier will only sell in lots of 10, 20, 30, or 40 pairs. Any orders for more than 50 or less than 10 will not be accepted.

(ii) The following quantity discounts apply on lot size orders:

Lot size	:	10	20	30	40
Discount (per cent)	:	5	5	10	20

(iii) For each order placed, the store incurs a fixed cost of Rs 20. In addition, the supplier charges an average amount of Rs 50 per order to cover transportation costs, insurance, packaging and so on, irrespective of the amount ordered.

(iv) Due to large in-process inventories, the store will carry no more than 40 pairs of shoes in inventory at the end of any one month. Carrying charges are Rs 5 per pair per month, based on the end of month inventory. It is desirable to have both incoming and outgoing seasonal inventory at zero.Find an ordering policy which will minimize total seasonal costs.

18. A man is engaged in buying and selling identical items, each of which requires considerable storage space. The buying and selling prices are indicated in the table below. He operates from a warehouse which has a capacity of 500 items. He can order on the 15th of each month, for delivery on the first day of the following month. During a month, he can also sell any amount up to his total stock on hand.

	January	*February*	*March*
Cost Price (Rs)	150	155	165
Sales Price (Rs)	165	165	185

If he starts the year with 200 items in stock, how much should he plan to purchase and sell each month, in order to maximize his profit for the first quarter of the year?

19. A pharmaceutical company has ten medical representatives working in three sales areas. The profitability for each representative in three sales areas is as follows:

No. of Representatives		: *0*	*1*	*2*	*3*	*4*	*5*	*6*	*7*	*8*
Profitability	Area 1	:15	22	30	38	45	48	54	60	65
(Rs '000)	Area 2	:26	35	40	46	55	62	70	76	83
	Area 3	:30	38	44	50	60	65	72	80	85

Determine the optimum allocation of medical representatives in order to maximize the profits. What will be the optimum allocation to the number of representatives available at present is only six?

20. A company has three media A, B and C available for advertising its product. The data collected over the past years about the relationship between the sales and frequency of advertisement in the different media is as follows:

Frequency/Month	*Estimated* *A*	*Sale (units) per Month* *B*	*C*
1	125	180	300
2	225	290	350
3	260	340	450
4	300	370	500

The cost of advertisement is Rs 5,000 in medium A, Rs 10,000 in medium B and Rs 20,000 in medium C. The total budget allocated for advertising the product is Rs 40,000. Determine the optimal combination of advertising media and frequency.

21. Neema-Chem manufacturers use cupric chloride as the basic material for the production of copper complex. For a smooth functioning of its production schedules, the company may have

enough stock every month of its basic material. The purchase price p_n and the demand d_n forecast for the next 6 months by the management is given below:

Month (n)	:	1	2	3	4	5	6
Purchase price (p_n)	:	11	18	13	17	20	10
Demand (d_n)	:	8	5	3	2	7	4

Due to limited space, the warehouse cannot carry more than 9 units of the basic material. The basic material is bought at the beginning of each month. When the initial stock is 2 units and the final stock is required to be zero, the company wishes to find an ordering policy for the next 6 months so as to minimize the total purchase cost.

22. An enterprising researcher believes that he has developed a system for winning a popular Las Vegas game. His colleagues do not believe that this is possible, so they have made a large bet with him. They bet that, starting with three chips, he will not have at least five chips after three plays of the game. Each play of the game involves betting any desired number of available chips and then either winning or losing this number of chips. He believes that his system will give him a probability of 2/3 of winning a given play of the game. Find his optimal policy regarding how many chips to bet, if any, at each of the three plays of the game in order to maximize the probability of winning his bet with his colleagues.

3

Analysis of Project Scheduling

INTRODUCTION

All conventional scheduling systems are based on fixed time estimate of activities, which, in actual situations, may not be so. In the event of time estimates of jobs, not predicted accurately, there would be uncertainty and risk. It is, therefore, natural for the project-in-charge to know the risks involved ad extent of uncertainty associated *with* the project. The main positive feature of PERT network over CPM network is the former's ability to provide help for management decisions under conditions of uncertainty. The concept provides probability that a certain project would be completed by the given date.

Based on the spread of b–a, the activities may be called deterministic (b – a spread is small) and variable (b – a spread is fairly large). Most indus-trial activities are deterministic in nature, while activities of R and D projects are variable in nature. As discussed above, three-time estimates (a, b, and m) are used for variable activities.

For the calculations of the probability of project completion in a given time, following points should be kept in mind:

(i) In majority of the situations, the data on probability of occurrence of an activity *vs.* durations will conform to β-distribution (Fig. 1.7).

(ii) The expected time divides the area under the curve into two parts, *i.e.,* the probability of completing an activity within its expected duration is 50%.

(iii) The expected time (t_e) of the activity is located at one-third of the distance between *m* and midrange (a + b)/2 away from most likely time.

(iv) Standard deviation (σ) and variance (σ^2) which are the measures of spread of data and respectively equal to (b – a)/6 and help in determining the probability of achieving the target completion date of the project or any stage of the project.

(v) Although, the expected time (t_e) of each activity independently has β-distribution, yet the completion time T has a normal distribution with mean T_{ep} and variance V_t. This statement can be clarified as under:

Suppose a project consists of *n* independent critical tasks with expected times of t_{e1}, t_{e2} ..., t_{en}. Each of these tasks, though expected to follow β-distribution, their total time (project duration) T_{ep}, is expected to follow normal distribution with mean T_{cp} and variance of critical path V_t.

$$T_{cp} = t_{el} + t_{e2} + ..., t_{en}$$

$$V_t = V_{t1} + V_{t2} + ..., V_{tn}$$

The conclusion is based on the concept of central limit theorem.

Using the above mentioned principles, the probability of completion of project can be established as under:

Let t_1. t_2........t_n = the expected times of he activity on the critical path.

T_{cp} = Expected time of completion of the project.

$$= t_1 + t_2 + ... + t_n$$

T = scheduled time of the project

$$\sigma = (V_{t1} + V_{t2} +... + V_{tn})^{1/2}$$

Once T_{cp} and σ have been calculated, then from tables of normal curve, the probability corresponding to standard normal deviate (Z) can be read, Here

$$Z = (T - T_{cp})/\sigma$$

Activity duration time has to be arrived by the experienced team having full knowledge of the work to be done as the subsequent network analysis is based on activity duration only. The time, so worked out, is entered in the network.

Description of activity is given over the arrow and time is indicated below the arrow, *e.g.*, design machine over three weeks. Unit of time should be uniform throughout the network in terms of weeks, days, hours, months, etc.

Analysis of Activity Durations

Mere computation of event time is not sufficient, it is equally important to establish the date at which each activity should start and end to maintain:

(i) Earliest starting date (ES) which implies that the earliest time of an activity (TE) of the tail event, *i.e.,* $ES_{ij} = TE_i$

(ii) Earliest finishing date which equals the earliest event time of tail event plus the duration of the activity emanating from the tail activity,

i.e., $EF_{ij} = TE_i + t_{ij}$

or $EF_{ij} = ES_j + t_{ij}$

(iii) Latest finishing date which is the latest event time of the head event, *i.e.,* $LF_{ij} = TL_{ij}$.

(iv) Latest starting date which is the latest finishing time minus activity duration *i.e.,* $LS_{ij} = LF_{ij} - t_{ij}$.

CALCULATION OF EARLIEST EXPECTED EVENT TIME (TE)

The earliest expected time of an event is the earliest time by which the event can be completed. Since an event cannot occur earlier than this point of time, it is termed as *Earliest Time Event.* Once the elapsed time (t_e) for each activity is worked out on the network, *Earliest Expected Time* for each event TE (the earliest time when an event may be expected to occur) is calculated by forward pass computation from the beginning event by summing up the activity times through the longer path on the network. When two or more activities constrain a single event, *TE* is calculated along each path and the longest time is chosen as the *TE* of the given event. It is because an event can occur only after all the preceding activities are completed. *TE* is written in square box above the event. *TE* for the end event determines the total project duration, *i.e.,* the time needed to finish the whole project.

Initial project event can be started arbitrarily with an occurrence time of zero. Based on zero starting time, the earliest expected time for the next event is computed by adding duration (t_e) of the activity path leading to this event. In other words, head and tail event times are treated as the boundaries between which the activities can move. But when two or more activities are joining a particular event, then *TE* is calculated by taking the higher or largest of all the time values for the merging activities leading to an event. In other words,

(i) The earliest occurrence of initial event is taken as zero, $TE = 0$.

(ii) Each activity begins as its predecessor event occurs = $TE + t_e$.

(iii) The earliest merge event time is the largest of earliest finish time of merging activities.

TE calculations can be explained with the help of the following illustration (Fig. 1).

In the above diagram (Fig. 1.8), event 1 has zero time. It takes 4 weeks on path 1-2 for an activity to be completed. Hence, *TE* for event 2 becomes 4. For activity 1-3, it takes 3 weeks and as such *TE* for event 3 is 3. When we want to reach event 4, then there are options available, *i.e.,* 3-4 and 2-4. Let us work out *TE* for each path as under

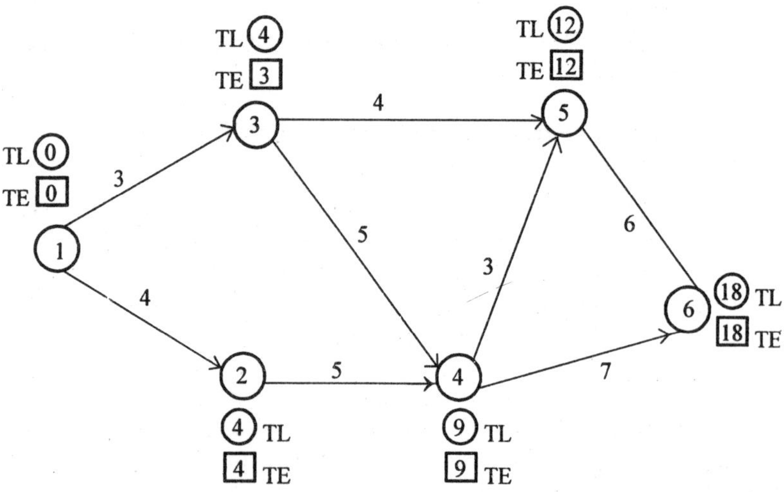

Fig 1 : Illustration for TE and TL Calculations.

Path 3-4 *TE*-3 + 5 = 3 + 5 = 8

Path 2-4 *TE*-2 + 5 = 4 + 5 = 9.

The highest/largest value in the two paths is 9 and as such *TE*-4 would be 9.

As regards event number 5, we can reach at event 5 by adopting paths 3-5 and 4-5.

Path 3-5 *TE*-3 + 4 = 3 + 4 = 7

Path 4-5 *TE*-4 + 3 = 9 + 3 = 12

Thus *TE*-5 = 12

As regards event 6, we can reach at event 6 by 4-6 and 5-6

Path 4-6 *TE*-4 + 7 = 9 + 7 = 16

Path 5-6 *TE*-5 + 6 = 12 + 6 = 18

Thus *TE*-6 =18.

LASTEST ALLOWABLE EVENT TIME (TL)

The next step is the determination of the latest allowable event time (TL), which is the latest time when an event can occur without creating any expected delay in the completion of the end event. The *TL* for the end event is set as equal to *TE* of the end event or equal to any pre-determined, specified, or directed time. The significance of the latest allowable time lies in the fact that if any of the events is delayed beyond the permissible *TL* values, it is bound to affect project completion in the desired period. In other words, *TL* values act as a warning signal to achieve the event as expected. Otherwise, it would lead to slippages in terms of time, cost, as well as performance. The *TL* for any given event "*X*" is thus calculated by Backward Pass computations from the end event as below:

Generally, value of end event (*TE*) is set equal to *TL*. The values of *TL* are worked out by subtraction process, *i.e.,* subtract from *TL* of each immediately following event the t_e leading to that event from event X, *i.e., TL* of an event is *TL* of its successor event minus the duration of activity joining the two events. If two alternative paths back to the event, different results would be obtained. The smallest out of the different results' values is chosen. This value also represents the longest backward time path from the end event.

The *TL* values thus calculated can be represented in small circles placed near the events (Fig. 1.8). The *TL* for the start event would then indicate the latest time by which the job can be started without causing the end event to slip beyond the pre-determined target time.

It is from the end event that we shall move backward to find out *TL* values for event 5, 4, 3, 2, and 1, respectively. These calculations are worked as shown in Table 1.

Table 1 : Calculation of TL Values

Event No.	Details	Values
6	Value is taken as that of its TE	18
5	TL 6–activity duration of 5-6	18–6 = 12
4	TL 6–activity duration of 4-6	18–7 = 11 Min. Value to be taken
	or	
	TL 5–activity duration of 4-5	12–3 = 9
3	TL 5–activity duration of 3-5	12–4 = 8 Min. Value to be taken
	or	
	TL 4–activity duration of 3-4	9–5 = 4
2	TL 4–activity duration of 2-4	9–5 = 4
1	TL 3–activity duration of 1-3	4–3 = 1 Min. Value to be taken
	or	
	TL 2–activity duration of 1-2	4–0 = 0

Slack or Float

Slack or float is the amount of time an event can be delayed beyond its *TE* without affecting the *TL* of the final event. Thus, it is equal to $s = TL - TE$. For any event, slack may be zero, positive or negative.

Zero slack ($TL = TE$) means that exactly enough time has been allowed for the activity and spare time is not available, *i.e.,* job would be on time.

Positive slack ($TL > TE$) means that there is enough time needed to finish the job. If slack for the end event is positive, *i.e.,* directed time is later than the computed *TE* for the end event, the project would be ahead of schedule. A relatively large positive slack identifies the network path which will allow the reduction of resources in its share without causing any delay in the completion of the project as a whole. These spare resources can be transferred from such paths to other paths requiring resources. This results in reduction of the total duration of the overall project.

Negative slack ($TL < TE$) means that sufficient time has not been allowed for accomplishment of an event and indicates "apparent trouble". Where negative slack occurs, attention should be focussed to these areas, most warranting the action to reduce the time required to complete the job.

Following the determination of the critical path, the floats for the noncritical activities must be computed. Naturally, a critical activity must have zero float. In fact, this is the main reason, it is critical. Before showing how floats are determined, it is necessary to define two new times that are associated with each activity. These are the latest start (LS) and the earliest completion (EC) times, which are defined for activity (i, j) by

$$LS_{ij} = LC_j - D_{ij}$$
$$EC_{ij} = ES_i + D_{ij}$$
$$D_{ij} = \text{Duration of activity } i, j.$$

Various floats are, total float (TF), free float (FF) and the independent float (IF).

Total Float (TF_{ij})

It is the duration by which an activity can be delayed without affecting the project schedule, *i.e.*, excess available time over its duration. Therefore, total float is the maximum leeway available to an activity when all preceding activities occur at the earliest possible time and all succeeding activities occur at the latest possible time. Therefore, total float of an activity can be obtained as follows:

Total Float TF_{ij} = maximum time available to perform the activity –

(LC_j ES_i) – the duration of the activity (D_{ij})

or $$TF_{ij} = LC_j - ES_i - D_{ij} = LC_j - EC_{ij} = LS_{ij} - ES_i$$

Free Float (FF_{ij})

The free float is defined by assuming that all the activities start as early as possible. This refers to that portion of float within which an activity can be manipulated without affecting the float of succeeding activities. Free float results when preceding activities occur at the earliest event times and all succeeding activities also occur at the earliest event times. Thus, free float is the excess of available time ($ES_j - ES_i$ over its duration D_{ij}) *i.e.,*

$$FF_{ij} = ES_j - ES_i - D_{ij}$$

Independent Float

This refers to that portion of total float within which an activity can be delayed for start without affecting float of the preceding activities. This is the leeway available to an activity when all preceding activities occur at the latest event times and all succeeding activities occur at the earliest event times. Independent float can be obtained as follows:

Independent float = Earliest event timer of head event – latest event time of tail event – activity duration

or $$IF_i = TE_j - TL_i - t_{ij}.$$

Alternatively, independent float can be obtained by subtracting the tail event float from free float. If IF_i is negative, it is to be taken as zero.

Critical Path

The application of PERT-CPM should ultimately yield a schedule specifying the start and completion dates of each activity. The network diagram represents the first step towards achieving that goal. Due to inter-relation among the various activities, the determination of the start and completion time requires special computation. The calculations are performed directly on the network diagram using simple arithmetic. The end result is to classify the activities of the project as *critical or non-critical*. An activity is said to be critical, if a delay in its start will cause a delay in the completion date of the entire project. A non-critical activity is such that the time between its earliest start and its latest completion dates (as allowed by the project) is longer than its actual duration. In this case, the non-critical activity is said to have slack or float time.

Floats suggest the leeway available and its knowledge provides the flexibility from scheduling point of view. The network, therefore, may contain a path(s) which have no leeway or have zero float. Such a path(s) is (are) called *critical path(s)* which have no leeway or have zero float. For example, in the above network, 1-2, 2-3, 3-4, 4-6, 6-7, 7-8 comprises zero float and, thus, is a critical path.

In other words, the critical path has the least algebraic slack and it determines the minimum or earliest time required for completion of the overall project as the sequence of the activities on this path imparts the most rigorous time constraint on attainment of the end event. If any time is to be saved on the overall project, it should be saved on the activities, which fall on the critical path.

Since all the network paths, other than the *critical path*, are shorter, they have some amount of slack or free time. The identification of the *critical path* from the slack or *non-critical paths*, would indicate possibilities of diverting resources from activities on *non-critical paths* (which have some slack) to activities on the *critical path*, thereby, leading to reduction in the total project duration. In this way, the project can be brought within a desired schedule of completion by the application of a given measure of resources. Indiscriminate expediting, which may be purposeless and expensive, can, thus, be avoided, and selective use of the available resources on really worthwhile tasks can be ensured.

The method of determining a *critical path* calculations include two phases, *i.e., forward pass* and *backward pass*. In forward pass, the calculations begin from start node and move to the end node. At each node, a number is computed representing the earliest occurrence time of the corresponding event. These numbers are shown in squares(❐). In backward pass, calculations begin from the "end" node and move to the "start" node. The number computed at each node (shown in triangles) represents the latest occurrence time of the corresponding event.

The objective of critical path analysis is to estimate the total project duration and to assign starting and finishing times to all activities involved in the project. This helps in checking actual progress against the scheduled duration of the project.

The duration of individual activities may be uniquely determined (in case of CPM) or may involve the three time estimates (in case of PERT) out of which the expected duration of an activity is computed. Having done this, the following factors should be known to prepare project scheduling.

(i) Total completion time of the project.

(ii) Earlier and latest start time of each activity.

(iii) Float for each activity, *i.e.*, the amount of time by which the completion of an activity can be delayed without delaying the total project completion time.

(iv) Critical activities and critical path.

Consider the following notations for the purpose of calculating various times of events and activities.

E_i = earliest occurrence time of an event, I. It is the earliest time at which an event can occur without affecting the total project time.

L_i = latest occurrence time of an event, I. It is the latest time at which an event can occur without affecting the total project time.

ES_{ij} = earliest start time for an activity (i, j). It is the time at which the activity can start without affecting the total project time.

LS_{ij} = latest start time for an activity (i, j). It is the latest possible time by which an activity must start without affecting the total project time.

EF_{ij} = earliest finish time for an activity (i, j). It is the earliest possible time at which an activity can finish without affecting the total project time.

LF_{ij} = latest finish time for an activity (i, j). It is the latest time by which an activity must get completed without delaying the project completion.

t_{ij} = duration of an activity (i, j).

It has already been mentioned that in a network diagram there should only be one initial event and one end event, while other events are numbered consecutively with integer 1, 2, ..., n, such that i < j, for any two events i and j connected by an activity which starts at i and finishes at j.

For calculating the above mentioned times, we shall discuss two methods namely *forward pass* method and *backward pass* method.

Forward Pass Method

Let ES_i = Earliest start time of all the activities emanating from event *i*.

= Earliest occurrence time of event *i*.

If *i* = 1 is the "start" event, then conventionally, for the critical path calculations, $ES_1 = 0$.

Let D_{ij} be the duration of activity (*i, j*).

The forward pass calculations are thus obtained from the formula,

$ES_i = \max (ES_i + D_{ij})$, for all (*i, j*) activities defined where $ES_1 = 0$. Thus, to compute ES_j for event j, ES_i for the tail events of all the incoming activities (i, j) must be computed first.

The square above event 1. Since there is only one incoming activity (1, 2) to event 1 with $D_{12} = 7$.

$$ES_2 = ES_1 + D_{12} = 0 + 7 = 7$$

Similarly,

$$ES_3 = ES_2 + D_{23} = 7 + 7 = 14$$
$$ES_4 = ES_3 + D_{34} = 14 + 2 = 16$$
$$ES_5 = ES_3 + D_{35} = 14 + 7 = 21$$
$$ES_6 = ES_4 + D_{46} = 16 + 6 = 22$$

Now the event 7 has 3 incoming activities, (3-7), (5-7), and (6-7), we have

$$ES_7 = \max_{i = 3, 5, 6} (ES_i + D_{i7})$$
$$= \max (14 + 5, 21 + 8, 22 + 8) = 30$$
$$ES_8 = ES_7 + D_{78} = 30 + 4 = 34$$

These calculations compute the forward pass.

Backward Pass Method

It starts from "end" event. The objective of this phase is to compute LC_i, the latest completion time for all the activities coming into event i. Thus, if $i = n$ is the "end" event, $LC_n = Es_n$ initiates the backward pass. In general, for any node i,

$$LC_i = \min [LC_j - D_{ij}] \text{ for all } (i, j) \text{ activities defined.}$$

The values of *LC* (entered in the triangles) are determined as follows:

$$LC_8 = ES_8 = 34$$
$$LC_7 = LC_8 - D_{78} = 34 - 4 = 30$$
$$LC_6 = LC_7 - D_{67} = 30 - 8 = 22$$
$$LC_5 = LC_7 - D_{57} = 30 - 8 = 22$$
$$LC_4 = LC_6 - D_{46} = 22 - 6 = 16$$
$$LC_3 = \min_{j = 4, 5, 7} (LC_j - D_{37})$$
$$= \min (16 - 2, 22 - 7, 30 - 5) = 14$$

$$LC_2 = LC_3 - D_{23} = 14 - 7 = 7$$
$$LC_1 = LC_2 - D_{12} = 7 - 7 = 0.$$

The backward pass calculations are now complete. The critical path activities can be identified by using the results of the forward and backward passes. An activity (i, j) lies on the critical path, if it satisfies the following three conditions:

$$ES_i = LC_i$$
$$ES_j = LC_j$$
$$ES_j - ES_i = LC_j - LC_i = D_{ij}.$$

These conditions actually indicate that there is no float or slack time between earliest start (completion) and the latest start (completion) of the critical activity. In the arrow diagram, these activities are characterized by the numbers in r and Δ being the same at each of the head and the tail events and that the difference between the number in r (or Δ) at the head event and the number in r (or Δ) at the tail event is equal to the duration of the activity.

The activities (1-2), (2-3), (3-4), (4-6), (6-7), and (7-8) define the critical path. Actually, the critical path represents the shortest duration needed to complete the project. Further, it should be noted that the critical path must form a chain of connected activities that spans the network form "start" and "end". (3, 5), (3, 6) and (4, 6) satisfy conditions (1) and (2) for critical activities but not condition (3).

Hence they are not critical. Notice also that the critical path must form a chain of connected activities that spans the network from "start" to "end".

UNCERTAIN ACTIVITY TIMES WITH PROJECT SCHEDULING

PERT was developed for the case when there is little past history on which to base estimates of activity durations (*e.g.,* R & D projects). Using PERT, time duration for each activity is no longer just a single time estimate, (*i.e.,* decision maker's best guess) but is now a random variable which is characterized by some probability distribution—usually a β (beta)-distribution. To estimate the parameters of the β-distribution (the mean and variance), the PERT model requires three time estimates for each activity. From these times a single value is estimated for future consideration. The three-time estimates required are as under.

(i) *Optimistic time* (t_o or a) This is the shortest possible time required to perform an activity, assuming that everything goes well.

(ii) *Pessimistic time* (t_p or b) This is the longest possible time required to perform an activity under extremely bad conditions. However, such conditions do not include acts of God like earthquakes, flood, etc.

(iii) Most likely time (t_m or m) This is the most likely time required to complete an activity. If the activity was repeated many times, then it is the duration that would occur most frequently (*i.e.*, model value).

The β-distribution is not necessarily symmetric, the degree of skewness depends on the location of t_m to t_o and t_p Thus, the range specified by the optimistic time(t_o) and pessimistic time (t) estimates is ,assumed to enclose every possible estimate of the duration of the activity. The most likely time (t_m) estimate may not coincide with the midpoint $(t_o + t_p)/2$ and may occur to its left or to its right. Because of these properties, it is justified to assume that the duration of each activity may follow Beta (β) distribution with its unimodal point occurring at t_m and its end points at t_o and t_p.

In Beta distribution the midpoint $(t_o + t_p)/2$ is assumed to weigh half as much as the most likely point (t_m). Thus, the expected or mean (t_o or μ value of the activity duration can be approximated as the arithmetic mean of $(t_o + t_p)/2$ and $2.t_m$. That is

$$\text{Expected activity time } (t_e) = \frac{(t_o + t_p)/2 + 2t_m}{3} = \frac{t_o + 4t_m + t_p}{6}$$

Estimating the variance is apparently based on an analogy to the normal distribution where 99 per cent of the area under normal curve is within ± 3σ from the mean or fall within the range approximately 6 standard deviation in length, therefore the interval (t_o, t_p) or range $(t_p - t_o)$ is assumed to enclose about 6 standard deviations of a symmetric distribution. Thus, if σ denotes the standard deviation, then

$$6\,\sigma \cong t_p - t_o \text{ or } \sigma = \frac{t_p - t_o}{6}$$

$$\text{Variance of activity time, } \sigma^2 = \left[\frac{1}{6}(t_p - t_o)\right]^2$$

Standard deviation, $\sigma = \sqrt{\text{Variance}}$.

FORMULATION OF INTEGER PROGRAMMING PROBLEM

The integer programming problem requires the following characteristics to be satisfied:

(i) A linear objective function

(ii) A set of linear resource constraints

(iii) Integer value constraints for all or some variables, and

(iv) Non-negativity constraints for the decision variables.

Using the mathematical notations as used in formulating a Linear Programming Problem, Integer Programming Problem can be formulated mathematically as

$$x_j \geq 0 \text{ for } j = 1, 2 \ldots\ldots\ldots n \qquad \ldots(1)$$

$$x_j \text{ are integer values for } j = 1, 2, \ldots\ldots s \qquad \ldots(2)$$

Therefore, the Integer Programming Problem may be considered as a special case of Linear Programming with an additional constraints of type (42 included in the Linear Programming Problem. In (2) when s = n (every variable is integer values), the problem is referred as pure Integer Programming Problem, otherwise it is called a *Mixed Integer Programming Problem.* Depending upon the areas of application, the objective function (1) may be either maximization or minimization. An Integer programming problem may include ≥ in-equalities and/or equalities as well.

One of the easiest ways of solving the IPP is to transform the problem into a form, which allows the application of LP. Business situations of mixed IPP are quite common, *e.g.*, a metal container manufacturing firm should cut various sizes of blanks from the metal sheet stock with minimum waste. One of the restrictions is that the whole blanks should be cut from sheets. It is simple to round off non-integer solutions in problems involving indivisible resources, especially if the rounding off is small relative to the values of the variables involved. However, such rounding off can result in solutions far from the optimal integer solution. Following example demonstrates the Integer Programming Problem case:

Although, several methods are available for solving the all-integer and mixed IPP, Gomory's cutting plane method and branch-and-bound algorithms are more popular.

GOMORY'S ALL INTEGER CUTTING PLANE METHOD

Historically, the first method for solving Integer Programming Problems was the cutting plane method. There are different approaches to cuts, but in this text, R.E. Gomory's fractional cut will be discussed, since it guarantees an optimal solution in a finite number of iterations.

A systematic procedure called Gomory's all integer algorithm will be discussed here for generating 'cuts' (additional linear constraints) so as to ensure an integer solution to the given LP problem in a finite number of steps. Gomory's algorithm has the following properties:

(i) *Additional linear constraints* never cut-off that portion of the original feasible solution space which contains a feasible integer solution to the original problem.

(ii) *Each new additional constraint* (or hyperplane) cuts-off the current non-integer optimal solution to the linear programming problem.

Method for Constructing Additional Constraint (Cut)

Gomory's method begins by solving the LP problem without taking into consideration the integer value requirement of the decision variables. If the solution so obtained is an integer, *i.e.,* all variables in the 'x_B' *column* (also called basis) of the simplex table assume non-negative integer values, the current solution is the optimal solution to the given ILP problem. But, if some of the basic variables do not have non-negative integer value, an additional linear constraint called the Gomory constraint (or cut) is generated. After having generated a linear constraint (or cutting plane), it is added to the bottom of the optimal simplex table so that the solution no longer remains feasible. The new problem is then solved by using the dual simplex method. If the optimal solution so obtained is again non-integer, another cutting plane is generated. The procedure is repeated until all basic variables assume non-negative integer values.

The procedure of developing a cut is discussed below. In the optimal simplex table, we select one of the rows, called *source row* for which basic variable is non-integer. The desired cut is developed by considering only fractional parts of the coefficients in source row. For this reason, such a cut is also referred to as *fractional cut.*

Suppose the basic variable x_r has the largest fractional value among all basic variables restricted to be integers. Then the rth constraint equation (row) from the simplex table can be rewritten as:

$$X_{Br} (= B_r) = 1.x_r + (a_{r1}x_1 + a_{r2}x_2 + \ldots)$$

$$= x_r + \sum_{j \neq r} a_{rj}x_j \qquad \ldots(1)$$

where x_j ($j = 1,2, 3,\ldots$) represents all the non-basic variables in the rth constraint (row) except the variables x_r and b_r ($= x_{Br}$) is the non-integer value of variable x_r. Let us decompose the coefficients of x_j, x_r. variables and x_{Br} into integer and non-negative fractional parts in Eqn. (1) as shown below:

$$[x_{Br}] + f_r = (1 + 0)\, x_r + \sum_{j \neq r} \{[a_{rj}] + f_{rj}\} x_j \qquad \ldots(2)$$

where $[x_{Br}]$ and $[a_{rj}]$ denote the largest integer obtained by truncating the fractional part from x_{Br} and a_{rj} respectively. Rearranging Eqn. (2) so that all the integer coefficients appear on the left-hand side, we get

$$f_r + \{[x_{Br}] - x_r - \sum_{j \neq r} [a_{rj}]x_j\} = \sum_{j \neq r} f_{rj}x_j \qquad \ldots(3)$$

where f_r is a strictly positive fraction ($0 < f_r. < 1$) while f_{rj} is a non-negative fraction ($0 \leq f_{rj} \leq 1$).

Since all the variables (including slacks) are required to assume integer values, terms in the bracket on the left hand side as well as on the right hand side must be non-negative numbers.

Since the left-hand side in Eqn. (3) is f_r plus a non-negative number, we may write it in the form of the following inequalities:

$$f_r \leq \sum_{j \neq r} f_{rj}x_j \qquad \ldots(4)$$

or

$$\sum_{j \neq r} f_{rj}x_j = f_r + s_g \text{ or } -f_r = s_g - \sum_{j \neq r} f_{rj}x_j \qquad \ldots(5)$$

where s_g is a non-negative slack variable and is called the *Gomory Slack Variable.*

Equation (5) represents Gomory's cutting plane constraint. When this new constraint is added to the bottom of optimal simplex table, it would create an additional row to the table along with a column for the nsw variable s_g.

ESTIMATION OF PROJECT COMPLETION TIME

As we are expecting a variability in the activity duration, the total project may not be completed exactly in time. Thus, it is necessary to calculate the probability of actually meeting the scheduled time to the project as well as activities.

where, T_e = expected completion time of the project,

Z = number of standard deviations the scheduled time or target date

lies away from the mean or expected date.

In order to find out the probability of completing the project in some given time, we shall consider only the expected length of the critical path and its variance. The expected time of the project can be calculated by adding the expected time of each activity lying on the critical path. Since it is assumed that the two activities are independent, therefore, variance of the critical path can be known by adding variance of critical activities.

But Z = 0.67, from normal distribution table is 0.2514. Thus, the probability of not meeting the due date is 25.14 per cent and of meeting the due date is 1 – 0.2514 = 0.7486 or 74.86 per cent.

Construction of the Time Chart and Resource Levelling

The end product of network calculations is the construction of the time chart (or schedule), which can easily be converted into a calender schedule for convenient use in the execution of the project.

The construction of the time chart must be made within limitations of the available resources, since it may not be possible to execute concurrent activities due to resource diversifications.

In this situation, total float for the non-critical activity becomes useful. By shifting a non-critical activity (back and forth) between its maximum allowable limits, it becomes possible to lower the maximum resources requirements.

In any case, even in the absence of limited resources, it is a common practice to use total floats to level resources over the duration of the entire project. In fact, this would mean a more steady work force compared to the case where the work force (and equipment) would vary drastically from one day to the next.

Cost Consideration in Project Scheduling

PERT-cost, developed in 1962, is an extension of PERT-time and makes use of cost-duration relationship, *i.e.,* it integrates time data. It incorporates both time and cost into network so that their trade-off can be calculated.

Here two time and cost estimates are indicated for each activity in the network. These are normal estimate and crash estimate. The normal time estimate is the same as expected time estimate (approximates the most likely time estimate in PERT) and the normal cost is the cost associated with this time. The crash time estimate is the one that would be required to speed up the work if no costs are spared in trying to reduce the activity time. The crash cost is the cost associated with doing the job on a crash basis in order to minimize the completion time. Crash costs may be in the form of over-time work, additional work force, or diversion of resources at a faster rate.

The point (D_n, C_n) represents the duration D_n and its associated cost C_n, if the activity is executed under normal conditions. The duration D_n can be compressed by increasing the allocated resoures, and hence, by increasing the direct costs. There is a limit, called crash time, beyond which no further reduction in the duration can be effected. At this point, any increase in resources will only increase the cost without reducing the duration.

The time cost relationship is assumed to be straight line for convenience, since it can be determined for each activity from the knowledge of the normal and crash points only, *i.e.*, (D_n, C_n) and (D_c, C_c). The weekly incremental cost I_c is the crash cost (C_c) minus the normal cost (C_n) divided by the normal time (D_n) minus the crash time (D_c). Thus, we have the relationship.

It is now possible to determine, how the incremental costs for each activity can be used to reduce the total project time at the least additional cost.

Under such conditions, the activity can be broken into a number of sub-activities, each corresponding to one of the linear segments. After defining the cost-time relaionships, the activities of the project are assigned their normal durations. The corresponding critical path is then computed and the associated costs are recorded. The next step is to consider reducing the duration of the project. This can be effected only if the duration of a critical activity is reduced and, thus, attention must be paid to such activities alone. To achieve a reduction in the duration at the least possible cost, one must compress as much as possible the critical activity having the smallest cost-time slopes.

The result of compressing an activity is a new schedule and perhaps a new critical path. The cost of the new schedule must be greater than that of the immediately preceding one. The new schedule must now be considered for compression by selecting the uncrashed critical activity with the least slope. The procedure is repeated until all critical activities are at their crash times. The final result of the calculations is cost-time curve for the various schedules and their corresponding costs.

It is logical to assume that as the duration of the project increases, the *indirect* costs. The sum of these two costs (direct + indirect) gives the total cost of the project. The optimum schedule corresponds to the *minmum* total cost. The detailed computations of such technique can be explained with the help of the following example:

UPDATE PROJECT

A project may not follow exactly the time schedule developed for it when it is actually executed. There are abound to be unexpected delays and difficulties in terms of delay in supply of materials, non-availability of some machine and/or breakdown of machines, non-availability of skilled manpower, natural calamity, etc. In such cases, it may be necessary to review the progress of network planning and scheduling. Such review of the progress helps in taking stock of the progress that has been made and making necessary change in the initial schedule in terms of time and resources required by incompleted activities in the project.

There is no rule about the time to go updating of the project network. The frequency of updating may be more when project duration is small because few slippages in detecting the progress will affect the project as a whole as the time for absorbing such slippages is less. But in case of large projects the frequency of updating may be less at the initial stages because a few initial slippages may be absorbed later in the project. However, to add dynamism to the nature and progress of work, updating may be carried out as frequently as economically possible.

Updating of the project can be done in two ways:

(i) Use the revised times estimate of incomplete activities and calculate from the initial event the earliest and latest completion time of each event in the usual manner to know the project completion time.

(ii) Change the complete work to zero duration and represent all the activities already finished by an arrow called the elapsed time arrow. Events in the revised network diagram are renumbered. The completion times of remaining activities are taken as the revised time.

Resource Allocation

When resources such as men, money, material, machinery, etc., are limited and conflicting demands are made for the same type of resources as project progresses, a systematic method for the allocation of resources becomes essential. The aim is to prevent the day-to-day fluctuation in the level of required resources and obtain a uniform resource requirement during the project duration. There are basically two approaches as discussed below, of maintaining the total amount of resources in use during the project performance which is nearly constant as possible through time and avoid major shifts in resource from one activity to another.

RESOURCE LEVELLING

The analysis aiming at stabilization of rate of resource utilization (relatively constant) by various activities at different times without changing the project duration is called resource levelling.

In order to stabilize the use of existing level of resources the total float of non-critical activities is used. By shifting a non-critical activity between its earliest start time and latest allowable time, project manager may be able to lower the maximum resource requirement.

The following two general rules are normally used in scheduling non-critical activities.

(i) If the total float of a non-critical activity is equal to its free float, then it can be scheduled anywhere between its earliest start and latest completion times.

(ii) If the total float of a non-critical activity is more than its free float, then its starting time can be delayed relative to its earliest start time by no more than the amount of its float without affecting the scheduling of its immediately succeeding activities.

Resource Smoothing

The analysis aiming to reduce peak demand for resources and reallocating among activities of a project in a manner so that the total project duration remains shortest is known as resource smoothing (or loading).

The procedure of carrying out resource smoothing can be summarized in the following steps.

Step 1: Calculate the earliest start and latest finish times of each activity and then draw a time scaled version (or squared) of the network. In this network critical path is drawn along a straight line and non- critical activities on both sides of this line. Resource requirement of each activity is given along the arrows.

Step 2: Draw the resource histogram by taking earliest start times or latest start times of activities on the x-axis and cumulative resource required on y-axis.

Step 3: Shift start time of non-critical activities first having largest float in order to smoothen the resources.

ADVANTAGES OF PERT

PERT's application facilitates to achieve lower costs, reduce project time and manpower needs, its utility and value is most apparent in those areas, which tend to be conducted only once or twice and are not of routine or repetitive type. PERT's advantage too lies in complex programmes having many tasks, interdependencies, and inter-relationships to be considered. Above all, its approach is restricted to new, untried activities and not easily likened to past experience.

Application

PERT is gaining popularity in industrial research and development, construction of bridges, buildings, power stations, dams, as also computer programming and installations, preparation of bids and proposals, shutting down and restarting of chemical plants, blast furnace, oil refineries, production control systems, pilot production runs, production facility change over for new models, etc., PERT has the potentiality to be widely employed in adver-tising, securities issues, introduction of new products, mergers or acquisitions and marketing plans.

PERT is one of the powerful aids in policy making, planning, decision making and implementation process.

CPM, at present, finds wider application by firms, universities, consultants, and industries, which have been responsible for its development on a competitive basis. CPM has been used successfully in commissioning and installation of plants as well as modification or overhauling of existing plants and equipments. CPM has been tested and approved in contracts, tenders and design specifications by architects, engineers, State High Way departments, and utility companies.

Manufacturing concerns resort to CPM for massive marketing research, product launching, advertisement campaigns, product change, sales promotion, office procedures and a variety of other type of projects.

The essence of CPM lies in planning, analyzing, scheduling and controlling the project activities.

CPM, however, cannot be gainfully employed in situations where there are continuous activities like flow production for the sole reason that it can be applied only where the start and the completion of the task is already identified.

Again like PERT, CPM is not a cure for all the ills, at the same time, it is very effective to explore the real situations for immediate corrective measures.

PROJECT CONTROL

An important use of the arrow diagram occurs during the execution phase of the project and thus, arrow diagram should not be discarded soon after the schedule is developed. It is a rare case that the planning phase will develop a time schedule, which can be followed truly during the execution phase. In general, some of the jobs are delayed or expedited depending upon the actual work conditions. If such disturbances occur in the original plan, it calls for developing a new time schedule for the left out portion of the project.

The time schedule is used mainly to check if each activity is on time. The effect of delay in a certain activity on the remaining part of the project can best be identified on the arrow diagram.

For example, as the project progresses, it is found that delay in some activities requires developing an altogether new schedule. How can the new schedule be obtained using the present arrow diagram? The immediate need is to update the arrow diagram by assigning zero values to the durations of the finished activities. Partly completed activities are assigned times equivalent to their unfinished parts. Changes in arrow diagram such as addition or deletion of any future activities should also be made.

Project Time-Cost Trade-off

In previous sections, we discussed how to schedule project activities in a logical sequence. The cost of resources consumed by activities were not taken into consideration. The project completion time can be reduced by reducing (crashing) the normal completion time of critical activities.

The reduction in normal time of completion will increase the total budget of the project. However, the decision-maker will always look for trade-off between total cost of project and total time required to complete it.

Project Crashing

Crashing is employed to reduce the project completion time by spending extra resources (cost). However, as shown in Fig. 1.11, beyond point A cost increases more quickly when time is reduced. Similarly, beyond point B, the time increases while the cost decreases. Since for technical reasons, time may not be reduced indefinitely, therefore, we call this limit as *crash point*. There is also a cost efficient duration called *normal point*.

T For simplicity, the relationship between normal-time and cost as well as crash-time and cost for an activity is assumed to be linear instead of being concave and/or discrete. Thus, the crash cost per unit of time can be estimated by computing the relative change in cost (cost slope) per unit change in time.

It is clear that one must be interested in the central region of the curve contained between points A and B. This helps in establishing a trade-off between time and cost, the direct cost of completing an activity per unit of time.

Remark : Crashing an activity means performing it in the shortest technically possible time by allocating to it necessary resources.

TIME-COST TRADE-OFF PROCEDURE

The method of establishing time-cost trade-off for the completion of a project can be summarized as follows:

Step 1: Determine the normal project completion time and associated critical path for the following two cases:

(i) When all critical activities are completed with their normal time. This provides the starting point for crashing analysis.

(ii) When all critical activities are crashed. This provides the stopping point for crashing analysis.

Step 2: Identify critical activities and compute the cost slope for each of these by using the relationship

The values of cost slope for critical activities indicate the direct extra cost required to execute an activity per unit of time.

Step 3: For reducing total project completion time, identify and crash an activity time on the critical path with lowest cost slope value to the point where

(i) another path in the network becomes critical, or

(ii) the activity has been crash to its lowest possible time.

Step 4: If the critical path under crashing is still critical, return to Step 3. However, if due to crashing of an activity time in Step 3, other path(s) in the network also become critical, then identify and crash the activity(s) on the critical path(s) with the minimum joint cost slope.

Step 5: Terminate the procedure when each critical activity has been crashed to its lower possible time. Determine total project cost (indirect cost plus direct cost) corresponding to different project durations.

BRANCH AND BOUND METHOD

Branch and Bound Method is a generic technique for solving a class of optimization problems. The Branch and Bound (B and B) method was originally developed by A.K. Land and A.G. Doig to solve the all integer, mixed integer and zero-one programming problem. The optimal solution is obtained by successive partitioning of the set of all possible solutions into two mutually exclusive and exhaustive subsets, and establishing limit for each set such that all solutions in the set are worse than the limit. The first part of the technique involving partitioning is known as *branching*, whereas the second part of establishing limit is called as *bounding*. Thus, in this method also, first a continuous integral programming problem is solved ignoring the integer valued condition. Branching signifies partitioning a current solution space into mutually exclusive subspaces. In the optimal solution, if some one of the variables, say x_j is not an integer, given integral programming problem is divided (partitioned) into two sub-problems. If x^*_j is the value of x_j in the optimal solution, then the following should be noted:

$$[x^*_j] < x_j < [x^*_j] + 1$$

Where $[x^*_j]$ is the integer part of the value of x_j.

Therefore, any feasible integer value of x_j should satisfy one of the following two conditions:

$$x_j \leq [x_j] \text{ and } x_j \geq [x_j] + 1$$

It may be noted that the above two constraints are mutually exclusive, *i.e.,* both cannot be true simultaneously and, therefore, both cannot be considered simultaneously in the integer programming problem. x_j is referred to as the *branching variable.*

Now two sub-integer programming problems are formed by adding these constraints separately to the continuous integer programming problem. Thus, an original problem is branched into two sub-problems. The branching process discards that portion of feasible region which involves no feasible integer solution. For example, in an optimal solution without the integer constraint, if $x_1 = 5/2$.

This obviously gives that $2 < x^*_1 < 3$, *i.e.,* in an integer valued solution, either $x_1 \leq 2$ or $x_1 \geq 3$. Thus, there will be no integer valued feasible solution in the space between $x_1 = 2$ and $x_1 = 3$.

The basics of branch and bounding algorithm can be more conveniently explained with the help of a numeric example.

AREAS OF APPLICATIONS OF INTEGER PROGRAMMING

The integer programming is considered as the technique to solve problems in linear programming structure involving the variables non-divisible. In practical business problems, there are many such situations, such as *capital budgeting, assignment problem, travel salesmen problems,* etc. Few such examples are discussed below:

Capital Budgeting Problems

This involves the situation of alternative decisions of investing or not investing. A mathematical model for capital budgeting is given by Weingartner.

If b_j = Net present value of *j*th investment proposal

x_j = An amount between 0 and 1 such that $x_j = 0$ when investment is not accepted and $x_j = 1$ when the investment is accepted.

C_{ij} = Net cash outflow required for the *j*th proposal in the *period i.*

Fixed Cost (or Charge) Problems

In certain cases, while undertaking a particular set of activities, the fixed costs (fixed charge or setup costs) are incurred. In such cases, the objective is the minimization of the total cost (sum of fixed and variable costs) associated with an activity:

Let us define the following decision variables

x_j = level of activity j

F_j = fixed cost associated with activity $x_j > 0$

c = variable cost associated with activity $x_j > 0$

Then the general fixed cost problem can be stated as:

$$\text{Minimize } Z = \sum_{j=1}^{n}(c_j x_j + F_j y_j)$$

subject to the constraints

$$\sum_{j=1}^{n} a_{ij}x_j \leq b_j; \qquad i = 1, 2, ..., m$$

$$x_j \leq My_j$$

or $\qquad x_j - My_j \leq 0; \; j = 1, 2,..., n$

$$x_j \geq \text{ for all j}$$

and $\qquad y_j = 0 \text{ or } 1 \text{ for all j}$

where the symbol M denotes a large number so that $x_j \leq M$.

Travelling Salesmen Problem

Suppose there are *n* cities with known distances between any two cities. A salesman has to start from his home city and visit each city once and then returns back to his starting point (say city-1). The objective is to minimize the total distance covered (or cost or time). The problem can be formulated as zero-one integer programming problem. The *Travelling Salesman Problem* can be formulated as following:

where d_{ij} = distance from city *i* to city *j*, and

1, if the salesman travels from city *i* to city *j* and city *j* is the *k*th city travelled

0, otherwise

(*i, j, k* arc integers varying from 1 to *n*)

The constraints involved would be as following:

(i) $\qquad k = 1, 2 \text{ ...... } n$
$\qquad i \neq j$

i.e., only one directed are may be assigned to a specific value of *k*.

(ii) $i = 1, 2, ...n$

i.e., only other city may be reached from a specific city *i.*

(iii) $j = 1, 2, ...n$

i.e., only one other city can initiate a directed arc to a specified city j.

(iv) for all j and k

$i \neq j, \quad r \neq j$

This constraint ensures that the complete travel will consist of a connected arc. It is given that the kth directed arc ends at some specific city j, the (k + 1)th directed arc must start at the same city j.

Assignment Problem

It refers to the situation where it is required to produce atleast N units (sizes) of a certain product on n different machines (or by n skilled workers), who are capable of producing any of the parts. However, each machine/worker requires different amount of time in producing a given product.

If

P_j = Number of products ordered of type j

x_{ij} = Number of products produced by machine/worker i

t_{ij} = Amount of time required for machine/worker *i* to produce a product of type *j*.

Since the objective of the company is to minimize the amount of time taken to produce all the required products, the problem can be formulated as below:

x_{ij} are non-negative integers for all i and j.

If the cost aspect is introduced in the objective function, the integer programming problem could be somewhat modified. If each skilled worker/machine is allowed a total of h_i hours to complete his work (*e.g.,* $h_i = 8$, for i = 1, 2,....m) and that the cost of having worker/machine i to produce a product of type j is C_{ij}, the problem then becomes

PLANT LOCATION PROBLEM

Suppose there are m possible sites at which plants could be located. Each of these plants produces a single commodity for n customers each with a minimum demand required for b_j units (j = 1, 2,..., n). The fixed set-up cost (expenses associated with constructing and operating a plant) for a plant in the rth location is f_i (i = 1, 2, ..., m). The production capacity for each plant is limited to a_i units. The unit transportation cost from plant j to customer j is c_{ij}. The problem is to locate the plants in such a way that the sum of the fixed set-up costs and transportation cost is minimized.

Let x_{ij} be the amount shipped from plant i to customer j, and y_i be the new variable associated with each of the possible plant locations, such that

$$y_i = \begin{cases} 1, \text{ if plant is located at the ith location} \\ 0, \text{ otherwise} \end{cases}$$

Constraints : (i) guarantee that each customer's demand is met. If all the shipping costs are positive, *i.e.,* there are no subsidized routes, then it never pays to send more than needed, or we replace the inequality sign by an equality. Inequality (ii) ensure that we do not ship from a plant which is not operating, u_i is the upper bound on the amount shipped which may be shipped from plant i and inequality (iii) restricts production from exceeding the limited capacity.

GOMORY'S MIXED-INTEGER CUTTING PLANE METHOD

In the previous section, it was important to note that the fractional cut assumes that all the variables, including slack and surplus, are integers. This means that the application of the fractional cut will yield no feasible integer solution unless all variables assume integer values. Simultaneously, for example, consider the constraint

$$\frac{1}{2}.x_1 + x_2 \leq \frac{11}{3}$$

or

$$\frac{1}{2}.x_1 + x_2 + s_1 = \frac{11}{3}$$

$$x_1, x_2, s_1 \geq 0 \text{ and integers.}$$

It may be noted that this equation can have a feasible integer solution in x_1 and x_2 only if s_1 is non-integer. This situation can be avoided in two ways:

1. The non-integer coefficients in the constraint equation can be removed by multiplying both sides of the constraint with a proper constant. For example, the given constraint above is multiplied by 6 to obtain: $3x_1 + 6x_2 \leq 22$. However, this type of conversion is possible only when magnitudes of the integer coefficients are small.
2. Use a special cut called *Gomory's mixed-integer* cut or simply mixed-cut where only a subset of variables may assume integer values and the remaining variables (including slack and surplus variables) remain continuous. The details of developing this cut are presented below.

METHOD FOR CONSTRUCTING ADDITIONAL CONSTRAINT (CUT)

Consider the following mixed integer programming problem:

Maximize $Z = c_1x_1 + c_2x_2 + \ldots + c_nx_n$

subject to the constraints

$$\begin{aligned} a_{11}x_1 + a_{12}x_2 + \ldots + a_{1n}x_n &= b_1 \\ a_{21}x_1 + a_{22}x_2 + \ldots + a_{2n}x_n &= b_2 \\ \vdots \quad & \quad \vdots \\ a_{m1}x_1 + a_{m2}x_2 + \ldots + a_{mn}x_n &= b_m \end{aligned}$$

and $\quad x_j$ are integers; $j = 1, 2, \ldots, k \; (k < n)$.

Suppose that the basic variable x_r is restricted to be integer and has a largest fractional value among all those basic variables which are restricted to take integer values. Then rewrite the rth constraint (row) from the optimal simplex table as follows [same as Eqn. (2)]

$$x_{Br} = x_r + \sum_{j \neq r} a_{rj}x_j \qquad \ldots(1)$$

where x_j represents all the non-basic variables in the rth row except variable x_r and x_{Br} is the non-integer value of variable x_r.

Decompose coefficients of x_j, x_r variables, and x_{Br} into integer and non-negative fractional parts as shown below:

$$x_{Br} = [x_{Br}] + f_r$$

and $\quad R_+ = \{j : a_{rj} \geq 0\}$ set of subscripts j (columns in simplex table) for which $a_{rj} > 0$

$R_- = \{j : a_{rj} < 0\}$ set of subscripts j (columns in simplex table) for which $a_{rj} < 0$

Then Eqn. (1) can be rewritten as:

$$[x_{Br}] + f_r = (1 + 0)\, x_r + 2\, a_r + \sum_{j \in R_+} a_{rj} x_j + \sum_{j \in R_-} a_{aj} x_j \qquad ...(2)$$

Rearrange the terms in Eqn.(2) so that all of the integer coefficients appear on the right-hand side. This gives

$$\sum_{j \in R_+} a_{rj} x_j + \sum_{j \in R_-} a_{aj} x_j = f_r + \{[x_{Br}] - x_r\} = f_r + I \qquad ...(3)$$

where f_r is a strictly positive fraction number (*i.e.*, $0 < f_r < 1$), and I is the integer value.

Since the terms in the bracket on the right-hand side of Eqn.(3) are integers, left-hand side in Eqn. (3) is either positive or negative according as $f_r + I$ is positive or negative.

Case 1 Let $f_r + I$ be positive. Then it must be $1 + f_r$, $+ 2 + f_r$, ... and then we shall have

$$\sum_{j \in R_+} a_{rj} x_j + \sum_{j \in R_-} a_{rj} x_j \geq f_r \qquad ...(4)$$

Since $a_{rj} \in R_-$ are non-positive and $x_j \geq 0$.

$$\sum_{j \in R_+} a_{rj} x_j + \sum_{j \in R_+} a_{rj} x_j + \sum_{j \in R_-} a_{rj} x_j$$

and hence $\sum_{j \in R_+} a_{rj} x_j \geq f_r$

Case 2 Let $f_r + I$ be negative. Then it must be f_r, $-1 + f_r$, $-2 + f_r$, ... and we shall have

$$\sum_{j \in R_-} a_{rj} x_j \leq \sum_{j \in R_+} a_{rj} x_j + \sum_{j \in R_-} a_{rj} x_j \leq -1 + f_r \qquad ...(5)$$

Multiplying both sides of Eqn. (10) by the negative number $\left(\frac{f_r}{f_r - 1}\right)$, we get

$$\left(\frac{f_r}{f_r - 1}\right), \sum_{j \in R_-} a_{rj} x_j \geq f_r \qquad ...(6)$$

Either of inequalities (4) and (6) holds, since in both the cases the left-hand side is non-negative and one of these is greater than or equal to f_r. Thus, any feasible solution to mixed-integer programming must satisfy the following inequality:

$$\sum_{j \in R_+} a_{rj} x_j + \left(\frac{f_r}{f_r - 1} \right) \sum_{j \in R_-} a_{rj} x_j \geq f_r \qquad \text{...(7)}$$

Inequality (7) is not satisfied by the optimal solution of the LP problem without integer requirement.

This is because by putting $x_j = 0$ for all j, the left-hand side becomes zero, and right-hand side becomes positive. Thus, inequality (7) defines a cut.

Adding a non-negative slack variable we can rewrite Eqn. (7) as

$$s_g = -f_r + \sum_{j \in R_+} a_{rj} x_j + \left(\frac{f_r}{f_r - 1} \right) \sum_{j \in R_-} a_{rj} x_j \qquad \text{...(8)}$$

Equation (8) represents the required *Gomory's cut.*

For generating the cut (8) it was assumed that the basic variable x_r should take integer value. But if one or more x_j, $j \in R$ are restricted to be integers, we can proceed as follows to improve a cut shown as Eqn. (7) to a better cut, *i.e.,* the coefficients a_{rj}, $j \in R_+$ and a_{rj} $\{f_r/(f_r - 1)\}$, $j \in R_-$ are desired to be as small as possible. The value of the coefficients of x_r can be increased or decreased by an integral amount in Eqn. (8) so as to get a term with smallest coefficients in Eqn. (7). Because in order to reduce the feasible region as much as possible through cutting planes, the coefficients of integer variable x_r must be as small as possible. The smallest positive coefficient for x_r. in Eqn. (3) is:

$$\left\{ f_{rj}; \frac{f_r}{1 - f_r}(1 - f_{rj}) \right\}$$

The smaller of the two coefficients would be considered to make the cut (8) penetrate deeper into the original feasible region. A cut is said to be *deep* if the intercepts of the hyperplane represented by a cut with the x-axis are larger. Obviously,

$$f_r \leq \frac{f_r}{1 - f_r}(1 - f_{rj}); \ f_{rj} \leq f_r$$

and

$$f_r > \frac{f_r}{1 - f_r}(1 - f_{rj}); \ f_{rj} > f_r$$

Thus, the new cut can be expressed as

$$s_g = -f_r + \sum_{j \in R} f_{rj}^{*} x_j \qquad ...(9)$$

; $a_{rj} \geq 0$ and x_j non-integer

; $a_{rj} < 0$ and x_j non-integer

; $f_{rj} \leq f_r$ and x_j integer

; $f_{rj} > f_r$ and x_j integer

Steps of Gomory's Mixed-Integer Programming Algorithm

Gomory's mixed-integer cutting plane method can be summarized in the following steps:

Step 1. Initialization Formulate the standard integer LP problem. Solve it by simplex method ignoring integer requirement of variables.

Step 2. Test of optimality (a) Examine the optimal solution. If all integer restricted basic variables have integer values, then terminate the procedure. The current optimal solution obtained in Step 1 is the optimal basic feasible solution to the integer LP problem.

(b) If all integer restricted basic variables are not integers, then go to Step 3.

Step 3. Generate cutting plane Choose a row r corresponding to a basic variable x_r which has largest fractional value f_r and generate a cutting plane as explained earlier in the form

$$s_g = -f_r + \sum_{j \in_+} a_{rj} x_j + \left(\frac{f_r}{f_r - 1}\right) \sum_{j \in_-} a_{rj} x_j$$

where $0 < f_r < 1$.

Step 4. Obtain the new solution Add the cutting plane generated in Step 3 to the bottom of the optimal simplex table as obtained in Step 1. Find a new optimal solution by using dual simplex method and return to Step 2. The process is repeated until all restricted basic variables are integers.

PROJECT MANANGEMENT

The main objective before starting any project is *to schedule the required activities in an efficient manner so as to complete it on or before a specified time limit at a minimum cost of its completion.* Hence, before starting any project, it is necessary to prepare a plan for scheduling and controlling the various activities (or tasks) involved with the given project. This will help in undertaking the project, possibly identifying bottlenecks and even discovering alternate work plan for the project. A project comprises a series of independent and/or interlinked activities each of which involves consumption of labour, materials, plant and machinery, and money. At the same time, the activities are interrelated in logical sequence, *i.e.*, some activities cannot start until others are completed. In general, a project is a one-time effort, *i.e.,* the sequence of activities may not be repeated in the future.

In order to execute a project efficiently and within the scheduled duration, critical activities consuming more time than others, should be identified and watched closely, whereas, at the same time, the activities allowing some delay in their scheduled completion, should be identified for more efficient use of resources. In the earlier time, there was hardly any planning in the scheduling of the project. The best known "planning" tool, used earlier, was the Gantt Bar Chart, which specifies the start and finish times for each activity on a horizontal time scale. It cannot determine the inter dependency between the activities which have a major role mainly in controlling the progress of the project. Present day projects are quite large and complex in nature and involve sophisticated designs, heavy capital investments over a period of time, a variety of materials, and variety of problems are faced during the execution of the project. As a result, more systematic and more effective planning techniques are required which should optimize the efficiency in executing the project, *i.e.,* effecting the utmost reduction in time required to complete the project while accounting for the economic feasibility of using available resources.

The techniques of operations research used for planning, scheduling and controlling large and complex projects are often referred as *network analysis, network planning* or *network planning* and *scheduling techniques*. All these techniques are based on the representation of the project as a network of activities. A network is a graphical plan consisting of a certain configuration of arrows and nodes for showing the logical sequence of various activities to be performed to achieve project objectives. It is perhaps the reason that the above nomenclature is given to these techniques. In this chapter, we shall discuss two of the well-known techniques, PERT and CPM belong to this family.

PERT (*Programme Evaluation and Review Technique*) was developed in 1956-58 by a research team to help in the planning and scheduling of the US Navy's Polaris Nuclear Submarine Missile project which involved thousands of activities. The objective of the team was to efficiently plan and develop the Polaris missile system. Since 1958, this technique has proved to be useful for all jobs or projects which have an element of uncertainty in the estimation of duration, as is the case with new types of projects the likes of which have never been taken up before.

At the same time but independently, CPM (*Critical Path Method*) was developed by E.I. DuPont company along with Remington Rand Corporation. The aim behind its development was to provide a technique for control of the maintenance of company' chemical plants. In course of time, use of CPM got extended to the field of cost and resource allocation.

In order to carry out the intended objectives and to have the optimum utilization of scarce resources, it has been established that the use of network Techniques helps in integrated cost planning, and financial and project scheduling on scientific basis. According to Drucker, organized study of "work" did not start until the beginning of 20th century. Taylor was the first one to have systematic observation and study of the work and he concluded that work is impersonal and objective, *i.e.*, work is a "task" and is something having a logic and needs analysis, synthesis and control. Therefore, the first step towards understanding work is to analyze it which involves:

(i) Identifying and analyzing the basic operations and,

(ii) Arranging the basic operations in a logical, balanced, and rational sequence. This will require the knowledge of production principles, which help in knowing the process of putting together individual operations into individual jobs, and individual jobs into production. This was well observed by *Gantt*, and thus, the *Gantt Chart* became the basis for today's Bar Charts/Bar Graphs/Calendar schedules/life cycle curves, etc.

Historical Perspective of Network Techniques

The Network techniques have their origin in the late fifties in U.S.A. The techniques were developed to facilitate planning, scheduling, and controlling the projects in an integrated manner with the aim to complete them within the constraints of given time and cost and required performance. According to United Nations Publication, *Analysis Through the Use of Network Techniques* is a managerial device which can satisfy a variety of needs such as system's design, planning, and control. Since older scheduling techniques did not enjoy much success in carrying out such projects, network is considered to be an important advancement in Project Management. There are two analytical techniques developed almost simultaneously (1956-1958) by two different groups for planning, scheduling and controlling the projects. These are the Critical Path Method (CPM) and the Project Evaluation and Review Technique (PERT).

CPM was first developed in 1957 by Morgan R. Walker of the Engineering Services Division of DuPont and James E. Kelley of Remington Rand. Walker and Kelley were concerned with the problem of improving scheduling techniques for such projects as building of a chemical Model Plant. Later on, it was applied for overhauling and maintenance of shutdown at DuPont Works.

At present, CPM is employed widely in different areas of industrial activities, particularly in construction industry. Through CPM, it is not essential to rush through and increase the cost of performing all the jobs, instead, crashing of a few activities alone can serve the purpose of expediting a project with good savings. PERT was developed in 1958 by U.S. Navy and a team of Management Consultants (Boose Allen and Hamilton) for scheduling the research and development activities for the Polaris Missile Programme. Since then, the use of PERT has spread rapidly throughout defense and space industry as well as in large industrial contracts in the field. Even small business houses found it increasingly necessary to develop PERT capability.

Potential Benefits of Network Techniques

Though, PERT and CPM techniques were developed independently, both have a common basis of optimization of resources for implementation of a project as per predetermined time, cost and performance. Important benefits of these techniques are as below:

1. A responsive tool is available for a long-term planning.
2. Inter-dependencies and problem areas, not definable under conventional planning techniques, are revealed.
3. A large data is available in a highly ordered fashion, and thus, the techniques introduce management "By Exception Principle" in the unknown area of planning and control.

4. A logical thinking device becomes available for preparation of project schedule and presenting a working logical model.
5. With the help of these techniques, the management can concentrate only on vital activities (10-20%), requiring most judicious allocation of resources.
6. Impact on total system, resulting from a change in original allocation of time and money, is projected.
7. A method for scientific allocation and utilization of scarce and limited resources comes in hand to achieve the goals within time and cost constraints.
8. An easy approach to reach the objectives is available.
9. The techniques act as an additional aid in rational decision-making, supervision, and control.
10. The technique is applicable to any size of organization.
11. A powerful means is made available to obtain trade-off between cost and time.
12. "One of a kind" programme as against a repetitive exercise.
13. The techniques help in predicting schedule slippage and cost over runs.
14. The techniques work as a diagnostic instrument in identifying weak areas for improvement.

Due to above listed advantages of *Network Techniques*, even the developing countries have also started relying on their application in almost all important sectors of economy, such as power, steel, irrigation, fertilizers, space programmes, etc. Recognizing the potential benefits of these techniques, the Administration Reform Commission, in 1965, stressed the need to use *Network Techniques* in systematic planning of approved projects. However, it may be noted that Network Techniques are not a universal medicine in themselves. These are one of the most important means for management to ensure efficiencies, effectiveness, and efficacy of projects.

The importance of these scientific techniques has been recognized to such an extent that big contracting firms have made it compulsory for their contractors to submit their bids accompanied by a network plan. Financial institutions and banks also instruct their prospective borrowers to append a network plan to their project feasibility reports.

PERT AND CPM DIFFERENCES

Over the years, PERT and CPM techniques have been developed in different environments. Although, objectives of the two techniques were initially extremely divergent, these are being used more profitably in conjunction with planning, design, scheduling, and control, all over the world. PERT and CPM are basically time oriented techniques, *i.e.,* both lead to determination of time schedule. However, CPM has an inherent capability of activity-cost optimization, while PERT on time estimation. Although, these techniques were developed independently by two different organizations, they are more or less similar and now referred to as the Network Techniques or Network Analysis or Critical Path Analysis or Project Scheduling Techniques. Inspite of their common features, style of functioning and close inter-action, the two methods can be clearly differentiated from each other as below:

1. PERT is an event-oriented technique, while CPM is an activity oriented one, *i.e.,* in the former, attention is focussed on starting and completion of event, rather than on the activities. Thus, CPM prepares network from activities, while PERT does it from events.
2. Originally, the time estimates for the activities in CPM are assumed deterministic, *i.e.,* single estimate of time, whereas probabilistic in PERT, *i.e.,* three times (most likely, pessimistic, and optimistic) formula was adopted for PERT analysis.
3. PERT allows for uncertainties in time, whereas CPM ignores chance element and employs only normal and crash cost time.
4. CPM stresses on cost concept, *i.e.,* deploying additional resources to shorten the duration of the job, whereas in PERT, more emphasis is on shortening and controlling project time on the understanding that reduction in time element would eventually lead to reduction in cost.
5. CPM is used mainly for construction programme, while PERT devotes its attention in areas like research and development programmes.
6. CPM relies on past experience, which is not taken into consideration by the PERT method.
7. The notations used in the two techniques are quite different. PERT makes use of notations like Network, event activity, slack, etc., whereas in CPM, notations like arrow diagram, node, job, float, etc., are inva-riably used.

Over the years, the distinction between the two techniques has diminished to a great extent due to various refinements, extension, change in formats, etc. This has given rise to a host of several other variations, namely, PEP (Programme Evaluation Procedure), SCANS (Scheduling and Controlling by Automated Network Systems), LCES (Least Cost Estimating and Scheduling), etc. Inspite of PERT and CPM extensions, the thumb rule is, when the time can be estimated fairly well and when costs can be calculated in advance. CPM is used, while the PERT is used in an extreme degree of uncertainty and control over time outweighing control over costs.

In other words differences of the PERT and CPM. Although there are no essential differences between PERT and CPM as both of them share in common the determination of a critical path and are based on the network representation of activities and their scheduling that determines the most critical activities to be controlled so as to meet the completion date of the project. However, following are the some of the other major differences.

PERT

1. It assumes a probability distribution for the duration of each activity. Thus completion time estimates for-all of the activities are needed.
2. To perform PERT analysis on a project, the emphasis is given on the completion of a task rather than the activities required to be performed to reach to a particular event or task. Thus, it is also called *an event-oriented technique.*
3. It is used for one-time projects involving activities of non-repetitive nature (*i.e.,* activities which may never have been performed before) in which time estimates are uncertain, such as redesigning an assembly line or installing a new information system.
4. It helps in identifying critical areas in a project so that necessary adjustment can be made to meet the scheduled completion date of .the project.

ANALYSIS OF CPM

1. This technique was developed in connection with a construction and maintenance project in which duration of each activity was known with certainty.

2. It is suitable for establishing a trade-off for optimum balancing between schedule time and cost of the project.
3. It is used for completion of projects involving activities of repetitive nature.

SIGNIFICANCE OF USING PERT/CPM

1. A network diagram helps translation of highly complex project into a set of simple and logical arranged activities and therefore,
   - help in the clarity of thoughts and actions.
   - help in clear and unambiguous communication developing from top to bottom and *vice-versa* among the people responsible for executing the project.
2. Detailed analysis of network help project incharge to peep into future because
   - difficulties and problems that can be reasonably expected to crop up during the course of execution, being foreseen well ahead of actual execution.
   - delays and hold-ups during course of execution get minimized. Corrective action can also be taken well in time.
3. Isolates activities which control the project completion and therefore, results in expeditious completion of the project.
4. Helps in the division of responsibilities and therefore, enhance effective coordination among different departments/agencies involved.
5. Helps in timely allocation of resources to various activities to achieve optimal utilization of resources.

Phases of Project Scheduling by Pert and CPM

The project scheduling by PERT and CPM consists of the following three phases:

1. Project Planning Phase, 2. Scheduling Phase, 3. Controlling Phase.

PROJECT PLANNING PHASE

In order to visualize the sequencing or precedence requirements of the activities in a project, it is helpful to draw a network diagram. For this following tips are adopted:

(i) Identify various activities (task or work elements) to be performed in the project, that is, develop a breakdown structure (WBS).

(ii) Determine requirement of resources such as men, materials, machines, money, etc., for carrying out activities listed above.

(iii) Assign responsibility for each work package. The work packages corresponds to the smallest work efforts defined in a project and form the set of elemental tasks which are the basis for planning, scheduling and controlling the project.

(iv) Allocate resources to work packages.

(v) Estimate cost and time at various levels of project completion.

(vi) Develop work performance criteria.

(vii) Establish control channels for project personnel.

(viii) Computing the time required for allocation of resources required for a project.

(ix) Breaking down the project into different activities.

(x) Determining the time estimates for these activities (allocation of resources for carrying out the work) for minimizing the total cost of the project.

(xi) Construction of Network (or arrow diagram) with each of its arcs (arrows) representing an activity.

SCHEDULING PHASE

Once all work packages (*i.e.,* tasks) have been identified and given unique names or identifiers, scheduling of the project, *i.e.,* when each of the activities required to be performed, is taken up. The various steps involved during this phase are listed below:

(i) Identify all people who will be responsible for each task.

(ii) Estimate the expected durations of each activity, taking into consideration the resources required for their execution in most economic manner.

(iii) Specify the interrelationship (*i.e.,* precedence relationship) among various activities.

(iv) Develop a network diagram showing the sequential interrelationship between various activities.

For this, tips such as; *what* is required to be done; *why* it must be done, can it be dispensed with; *how* to carry out the job; *what* must precede it; *what* has to follow; *what* can be done concurrently, may be followed.

(v) Based on these time estimates, calculate the total project duration, identify critical path; calculate floats; carry out resources smoothing (or levelling) exercise for critical (or scare) resources taking into account resource constraints (if any).

(vi) Studying the different jobs from arrow diagram (graphic representation) and suggesting improvements before the actual execution of the project.

(vii) Constructing a time chart showing the start and finish times for each activity as well as its relationship to other activities in the project and then preparing time bound targets for completion of various activities of the project.

(viii) Pinpointing the critical (in view of time) activities requiring special attention for on-time completion of the project.

(ix) Showing slack or float times, which can be used advantageously in the event of delaying or limited resources to be used effectively.

(x) Establishing requirements of men, materials, and money at different stages of the project.

Controlling Phase

Project control refers to evaluating actual progress (status) against the plan. If significant differences are observed, then the scheduling and resource allocation decisions are changed to update and revise the uncompleted part of the project. In other words, remedial (modifying planning) or reallocation of resources (cost minimization) measures are adopted in such cases.

The relationship among phases of project management which indicates how the actual task performance data is used to track deviations from the original plan and schedule. These adjustments are helpful to aggregate work packages into subsystems and track the progress of these subsystems as part of the reporting and review procedures. In this way, it is less likely that project manager will 'lose sight of the forest because of too many trees'.

(a) Using the arrow diagram and the time chart for making periodic progress reports.

(b) Defining areas of responsibility for different managers for timely execution of the project.

(c) Updating and analysis of the Network.

(d) If necessary, determining a new schedule for the remaining portion of the project.

OBJECTIVES OF THE NETWORK TECHNIQUES

The Network Technique constitutes an important step towards integrated management system. The Network technique, with its associated analysis helps, in obtaining the following objectives of an integrated management system:

(i) Detailed and integrated planning of the project tasks can be accomplished.

(ii) Realistic schedules can be developed.

(iii) Effective control can be exercised where periodic checking and evaluation of progress are required.

(iv) The progress of the project can be compared from time to time with the plan.

(v) Influence of the current progress can be projected on the completion time of the project, and thus, a corrective action can be taken in time.

(vi) It helps in making the optimum use of the available resources in men, money, and materials.

Other Planning and Control Techniques

A brief review of the evolution of traditional planning and control techniques will serve as a background for establishing the requirement for search of new methods and contribution of the network technique as an aid to the management. The important conventional planning and control methods are given below:

1. Flow Process Chart
2. Line of Balance Chart.
3. Gantt Chart
4. Milestone Chart

Flow Process Chart

These charts, developed by Gilbreths, represent the sequential relationship of activities or interrelationships between tasks. These charts provide an easy and effective means to communicate a plan with the help of a simple graphic presentation (Fig. 2). The flow charts are considered to be improvements over Gantt and Milestone charts, as they show relationship between activities and also certain events or milestones.

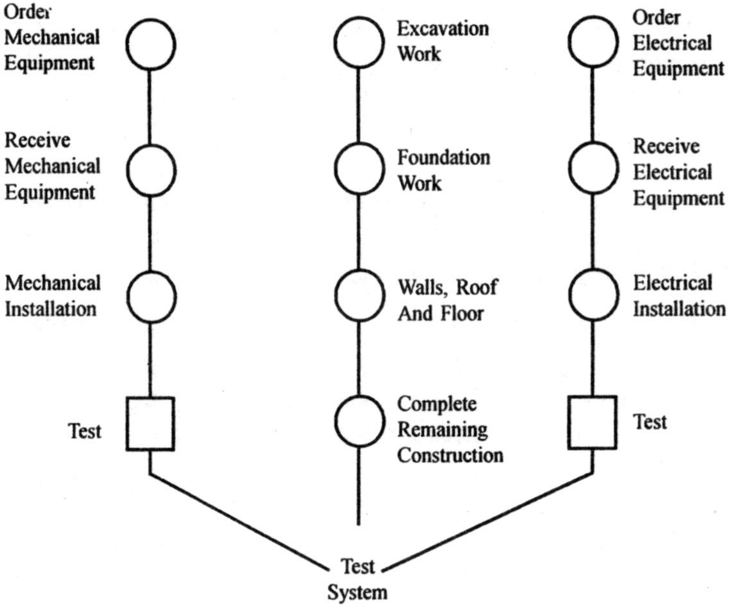

Fig. 2 : Flow Chart for a Building Project.

However, the flow charts reflect the following drawbacks:

(i) They fail in providing information regarding time durations of individual activities.

(ii) They do not help in determining the project schedule.

(iii) They do not help in predicting the completion time of the project.

The process charts can be used for the following:

(a) Work simplification.

(b) Technique improvement, and

(c) Procedure analysis.

Line of Balance Chart

It comprises a combination of the Process Chart and Gantt Chart, where the tasks are plotted on a time scale according to their sequential and lead time relationshps, leading to the final event.

This is used as a device for planning and monitoring progress of an order, project or a programme by a target date. The technique provides a control mechanism in monitoring production jobs, when quantity factor is involved. The steps involved in the use of Line of Balance Technique are:

(i) preparation of an objective chart depictiong the delivery schedule of the product and

(ii) the actual execution of the plan.

This concept can be explained with the help of the following example:

M/s *XY* Engineers, a manufacturer of water pumps has received an order of 3000 pumps from an automobile firm in Dec. 1999. The order is to be delivered from March 2000 according to the delivery schedule as given below:

Month	**March**	**April**	**May**	**June**	**July**	**August**	**Sept.**	**Oct.**
Quantity to be Delivered	200	300	200	300	500	700	500	300

The manufacturing process comprises of 9 distinct stages with time relationships as given below in Table-1.

Table 1 : Stages of Manufacturing Process

Stage	Decription	Time (Weeks) on Stage	Depends
1.	Purchase of castings	4	–
2.	Purchase of bar stocks	2	–
3.	Purchase of standard parts	1	–
4.	Purchase of made to order parts	8	–
5.	Machining of pump housings	4	1
6.	Manufacture of parts	2	2
7.	Assembly	2	3-6
8.	Testing	1	7
9.	Packing and dispatch	1	8

Gantt Chart

Prior to World War-I, the U.S. Army Ordinance Bureau indicated the need for a technique of planning and controlling the production of ordinance material. In pursuance of this, a consultant at Frankford Arsenal, Henry L. Gantt designed a simple Bar Chart (known as Gantt Bar Chart) displaying the schedule and the actual performance. In Gantt Chart, the steps, necessary to obtain a final work result, are worked out by projecting backward, step by step, from end result actions, with their timings and their sequence, showing graphical representation of work vs. time. The Gantt Chart shows bars drawn on time scale for each operation. Fig. 1.1 shows a typical Gantt Chart displaying the schedule for the construction of a small industry.

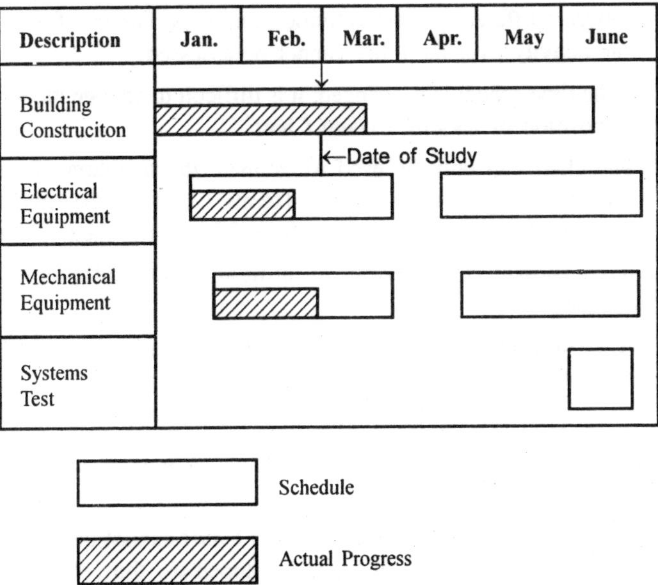

Fig. 3 : Gantt Chart Representing a Schedule for a Building Project.

In this Chart, periodically, the position of a bar could be shaded to indicate the progress made at the time of reporting. Fig. 3 shows the progress of the work on Feb. 28, according to which, foundation and building work are ahead of schedule, whereas the mechanical equipment work and electrical work are behind the schedule. Therefore, the project manager will become alert on mechanical equipment work and electrical work schedule.

In Gantt Chart, the time taken by an activity is represented by a horizontal line, the length of which being proportional to the duration of activity. As a rule, the time in the chart should flow from left to right and the activities be listed from top to bottom. The actual progress of the project can be represented by shaded bar, showing the amount of work completed. Thus, Gantt Chart not only represents the schedule, but can be considered as a good technique of reporting the actual progress and comparing it with the schedule.

Thus, Gantt Chart is an effective scheduling tool for simple and uncom-plicated projects, involving a minimum of coordination among the various tasks making up the entire project. However, the Gantt Chart has following deficiencies:

(i) It does not show the inter-dependencies or inter-relationships between different project activities, and thus, impact of delay of one activity over the other cannot be assessed.

(ii) If the time schedule is changed, it is difficult to change the length or position of bars of a Gantt Chart.

(iii) The chart can only be used to estimate the quantum of work, planned or completed only in broad terms, *i.e.,* at best the estimation could be done in terms of percentage (25-30%) of the total work.

(iv) It is not possible to achieve the scientific allocation and optimization of resources.

(v) Analytical integration of time, costs, work, and resources is quite difficult.

(vi) It is not possible to identify critical and non-critical areas of work.

(vii) Monitoring and control mechanism is neither effective nor a precise one.

Inspite of the above deficiencies, *Gantt Chart* cannot be discarded alto-gether. Various analysis techniques developed are, thus, elaborations of Gantt's work which have tried to eliminate the deficiencies and aim at providing a rational approach to project planning, implementation, and control.

Milestone Chart

The Milestone Chart, considered to be an improvement over the Gantt Chart, was developed around 1940. In principle, the Gantt Chart and Milestone Chart are quite similar, except that, the Milestone Chart has introduced the concept of events or intermediate milestones, which are inserted along the activity oriented horizontal bars. The identification made, with these positive milestones, make the reporting more definite.

However, Milestone Chart also does not show the inter-relationship among the various activities and, thus, finds limited predictive value.

If there is a schedule slippage in one of the milestones during various stages of the project execution, it is not possible to indicate the delay in the total project schedule.

DETERMINATION OF STAGE LEAD TIME

Lead times are the stage times needed to complete various activities of the project prior to the completion of the final assembly. Since the emphasis is on the major check points, detailed listing of activities is avoided and only major activities, such as procurement of materials, individual manufacturing processes, preparation of sub-assemblies, etc. are recorded. Following steps are required to establish stage lead times:

(i) An operation relationship chart involving drawing of a network for all activities is drawn first. Fig. 4 represents the network for the activities given below.

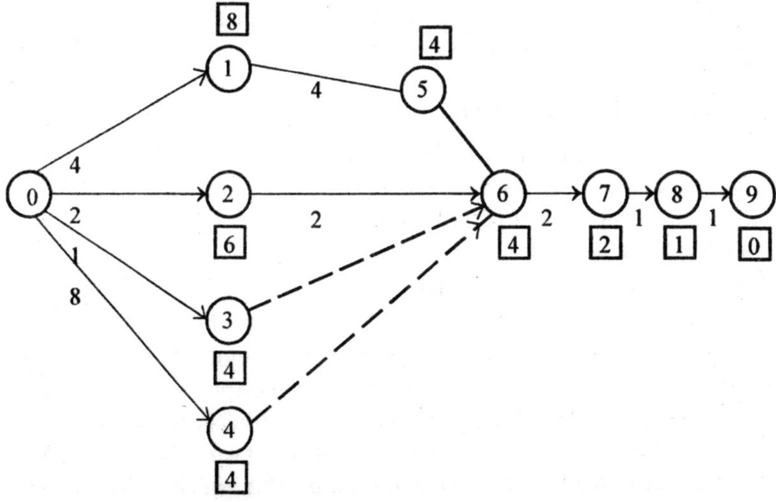

Fig. 1.3 : Network Showing the Stages of Manufacture.

(ii) Stage lead times are then calculated working back from final stage in question to the last stage. For example, a lead time for *i*th stage is the time lag between it (*i*th stage) and the final stage. For easier analysis, stage lead times are placed below the nodes and enclosed in square. The stages are next numbered according to their lead times, starting with the stage with the longest lead time.

(iii) An "operation plan chart" is prepared by placing each stage according to its lead times. This is a time chart starting with zero from final stage and extending from right to left. This chart represents all major and limiting stages in the production process. It is a time chart expressing graphically the lead time of each of the stages.

RECORDING OF PROGRESS OF WORK IN LINE OF BALANCE CHART

A Line of Balance Chart for each of the week of the manufacturing period is drawn before the commencement of the project and then the same is issued to the concerned sections/departments. The chart clearly indicates the progress of the work required at each stage to keep the manufacturing programme in line with the delivery schedule.

As the project is pursued, progress of each activity is plotted on Line of balance chart by shading out the corresponding column for the length equal to the actual progress.

Preparing a Progress Chart

A progress chart is prepared by taking cumulative values of deliveries and stage lead times. This chart will indicate the desired progress of work at each stage at a given period. Progress chart can be prepared by the following two methods:

(i) By converting the stage lead times into equivalent periods. In this case, equivalent periods (time until the period the progress is to be known) are calculated by adding the period to the stage lead times. In the above example, by week 8 (April end), 500 water pumps should be completed and despatched; but in order to maintain the delivery schedule, cumulative volume of work that should pass through the previous stage 8, must be 550, which is the volume of 8 + 1 weeks, 1 being the stage time between stages 8 and 9.

Similarly, 600 pumps must pass through stage 7 which is the volume at 8 + 2 weeks, 2 being the stage lead time between stages 7 and 9. From the above logic, progress can be prepared as given in the following Table:

Stage	Stage Lead Time	Equivalent Weeks	Period Months	Progress of the Work Through Stage at Week 8
1	8	8 + 8 = 16	4.00	1000
2	6	8 + 6 = 14	3.50	850
3	4	8 + 4 = 12	3.00	700
4	4	8 + 4 = 12	3.00	700
5	4	8 + 4 = 12	3.00	700
6	4	8 + 4 = 12	3.00	700
7	2	8 + 2 = 10	2.50	600
8	1	8 + 1 = 9	2.25	550
9	0	8 + 0 = 8	2.00	500

ANALYSIS OF PERFORMANCE

The gap between the top of the bar and the Line of Balance represents a delay, exception, or difficulty.

Line of Balance is an important technique for coordinating a large variety of production and purchase activities. It helps in the production line running by smooth flow of materials and prevents over accumulation of work-in-progress.

The stocks of materials may be reviewed on first day of the month and, if the actual progress is below the Line of Balance, control may be exercised to level it to the respective norms. And if the actual progress is above the Line of Balance, the schedules of activities may be adjusted to bring down to the respective norms.

With the latest application of Operations Research developed during War World-II, various prototype problems like inventory, queuing, sequencing, coordination, etc., had to be tackled by developing new techniques. PERT and CPM, which are the main network techniques, have tried to solve the problems under the head sequencing and coordination.

Preparing a "Line of Balance" Chart

An alternative to the equivalent week numbers is a line of balance progress which can be drawn as following:

(a) Vertical columns are drawn on X-axis (one column for each stage).

(b) The height of the column which is proportional to the production requirement of the stage (cumulative work required to flow at that stage) is obtained from projections across from the cumulative delivery schedule. The intersection of the column height and the line from the objective chart means the line of the corresponding operation or stage.

The progress chart so made for the above example (XY manufacturer) gives nine step-wise profile which indicates the desired progress of stages at the specified time.

Network Diagram Representation

The Network is a graphical diagram representing all dependencies and interrelationships amongst various activities of a work plan. It is also known as arrow diagram. The network is formed with the help of arrows and circles, which give technological relationships to the activities involved. The head of an arrow indicates the direction of progress in the project. Preparation of network needs thorough understanding of the network logic and basic elements of the network, which are activities, events, technological relationship, dummy activity, path, etc.

1. Activity

An activity is represented by an arrow (→). This is an element of a job, task or project having definite end. Thus, an activity is defined as recognizable part of a work in the project that consumes time and resources for its completion. The description of the activity is written above the arrow and the time required is written below the arrow, *i.e.* The "tail" end of an arrow shows "the beginning" point of an activity, whereas the "head" represents the "end" or completion of the activity, *i.e.*, beginning and end points of an activity are described by a preceding (tail) as well as the succeeding (head) events. Activities originating from a certain event cannot start until the activities terminating at the same event have been completed. Each activity must have a definite beginning and a definite end. The arrow is not a vector quantity and, thus, need not to be drawn on scale. It may be straight, long, short or bent, but not broken. The same activity can not be represented by two arrows. The arrows, indicating the activities, move in unidirection only, *i.e.*, from left to right and not *vice-versa*. An activity should be independent, *i.e.*, each arrow is used to represent exactly one and only one operation (activity) of a project, indicating which activity to precede, follow, or to take place simultaneously. However, a number of arrows may be used to represent different parts of the same operation. An activity indicates the tasks, such as preparation of design, erection of equipment, testing, procuring, etc.

2. Activity Relationship

A project is made up of a number of activities, which are interrelated. Three possible relationships are:

(a) Some activities may run parallel and are known as "concurrent" activities.

(b) Some activities depend upon completion of the others and are known as "succeeding" activities, and

(c) The activities, on which the succeeding activities depend, are known as "preceding" activities.

3. Dummy Activity

A dummy activity means that it shows only a dependence, but neither consumes time nor resources. It is used purely for convenience in drawing networks and is indicated by dotted line. Dummy activities serve two-fold purposes:

(i) Establish and maintain a realistic and perfect logical relationships between one activity and the other.

(ii) It keeps the events numbering in a systematic manner on the principle that the number of the head of arrow is always greater than at its tail. For example, in certain project, jobs *A* and *B* must precede *C*, while job *E* is preceded by job B only. Fig. 5(a) shows the incorrect relationship. The correct relationship using the dummy *D* is shown in Fig. 5 (b).

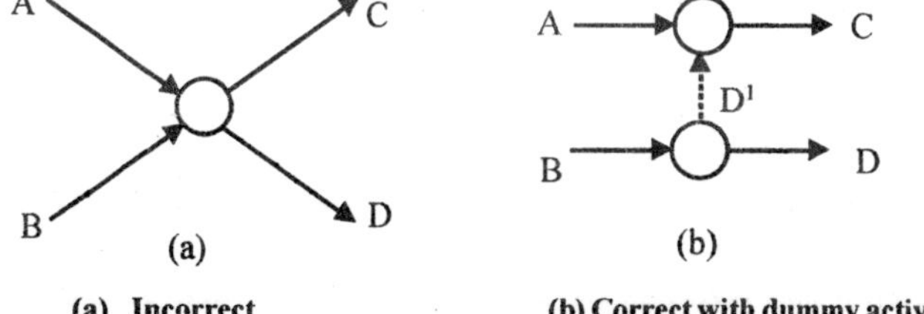

(a) Incorrect **(b) Correct with dummy activity**

D

Fig. 5 : Representation of Dummy Activity.

Above cases are illustrated in Fig. 6.

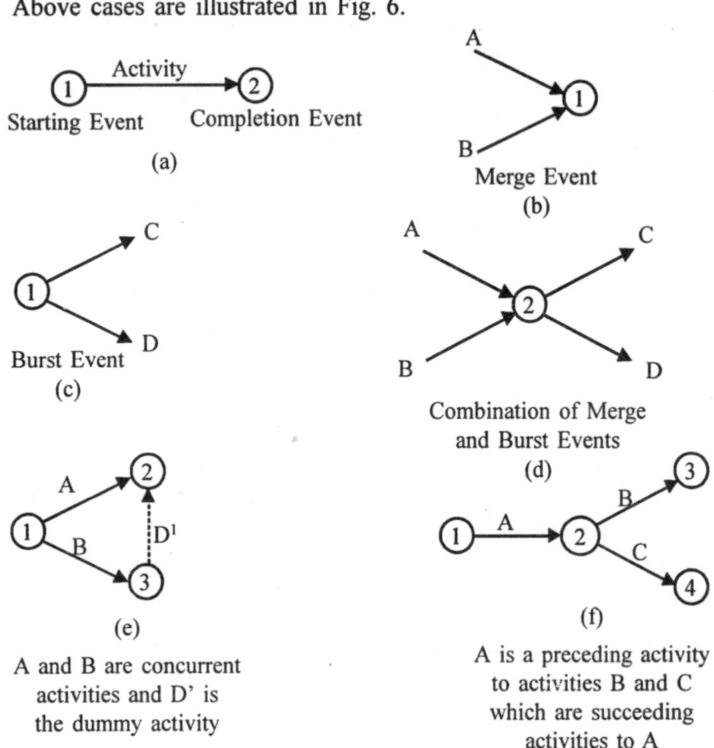

Fig. 6 : Representation of Various Activities and Events.

4. Loop Network

While drawing a network, loops in the reverse direction must be avoided. Loops occur by mistake from a duplicating event number or repetition of a particular activity or while the data is not being used accurately.

5. An Event

It is defined as an accomplishment occurring at an instantaneous point of time, *i.e.,* a point in time and not a passage of time, but consuming no time or resources by itself. In other words, an event represents a point in time that signifies the completion of some activities and the start of some new ones. An event is represented by a circle, rectangle, or square, etc. Thus, each arrow, representing the activity, must be bounded (circles, square, etc.). Events are usually described by words "such as" contract awarded, "project approved", "system tested", etc.

The start and end events, however, are reduced to a common event representing the completion of the first activity and the beginning of the next. Sometimes, a single event may represent the joint beginning of more than one activity (known as burst event) or the joint completion of more than one activity (merge event) or both (merge and burst event). A rectangle represents a milestone, *i.e.,* important events in the project. A hexagon is an indication of interface events, *i.e.,* events common to two or more than two networks. However, such events are outside the purview of project management.

6. A Path

A path, symbolized by a thick line (——) on the longest route, represents those critical activities, which have to be performed within the minimum stipulated period. A network can represent more than one critical path. Any delay in an activity on the critical path correspondingly affects delay in project implementation.

DETAILING THE NETWORK

The events in the network may be assigned a number to facilitate the description of activities, *i.e.,* event numbers, between which an activity lies, will represent that activity. The event numbering is carried out on sequential basis, *i.e.,* the head of an arrow should always bear a number higher than the one assigned at the tail of the arrow. Representation of an activity in numeric form has the advantages:

(a) An immediate identification of an activity can be made, since it is easier to pick out numbers than words at the network diagram.

(b) Reference is abbreviated.

(c) Sequence is immediately evident.

(d) Reference to sequential job is easier.

The activities must be defined in details but precisely. Activity description serves the purposes:

(a) Identifies the responsibility for performance of an activity.

(b) Determines the allocation of resources.

(c) Estimates the activity elapsed time.

Since dummies are as important as zeros in arithmetic, they should be introduced to serve the following purposes:

(a) To maintain the logic in network diagram.

(b) To maintain the uniqueness in the numbering system as every activity may have a distinct set of events (numbers) by which the activity can be identified. This is mainly applicable when a computer is employed for network analysis.

(c) To show the relationship between events, *i.e.,* when an activity has to be completed before another can begin.

Drawing of a network involves the following:

(1) Obtaining a precise statement of the project objectives.

(2) Determining the activities required to accomplish project objectives.

(3) Making use of paper about four times than the actual requirement.

(4) Drawing the diagram in pencil.

(5) Avoiding drawing arrows crossing each other.

(6) Drawing a skeleton network of the project (often with the aid of chart).

(7) Avoiding a wide variation in length of arrows.

(8) Drawing arrows horizontally on left-right principle except when dummies are introduced.

(9) Length of arrows must not be influenced with time estimation of the job.

(10) Introducing as many as dummies in the first draft, but weeding out those not required in the final draft.

(11) Soliciting and recording three time estimates for each activity.

(12) Numbering the events in the network.

NETWORK PREPARATION

The rules for constructing the network diagram may be summarized as:

Rule-1 : Each activity is represented by only one arrow in the network, *i.e.,* no single activity can be represented twice in the network or an event cannot occur twice. Thus, a path of activities cannot form a loop that returns to any event previously accomplished, *i.e.*, no event can depend for its completion upon the completion of a succeeding event. However, a case of one activity, being broken down into segments (each segment representing a separate task), should be differentiated. For example, laying down a pipe may be carried out in sections rather than as one job.

Rule-2 : No two activities can be identified by the same head and tail events. This type of situation may arise due to two or more activities being performed concurrently. In such a situation, a dummy activity is introduced.

Rule-3 : No event can occur until each activity preceding it has been finished.

Rule-4 : An activity, succeeding an event, cannot be started until that event has occurred.

Rule-5 : Each activity must terminate in an event.

Rule-6 : Time flows from left to right.

Rule-7 : Each activity on the network should be completed to reach the end objective.

Rule-8 : All individual tasks of a project should be visualized very clearly as to show on network. In order to develop a network with correct precedence relationship, following three questions should be answered in respect of each activity (arrow):

(i) What activities should precede (completed) the one being started *i.e.*, before this activity can begin?

(ii) What activities should follow this activity?

(iii) What activities can proceed concurrently?

Construction of Network

Once a project is accepted and approved by the competent authority, an important critical step in implementation process is to construct a network. Construction of a network mainly requires practice and different approaches have been suggested, the most common approach being to start with the first activity and draw the subsequent activities as per relationships. This, however, is a laborious method as it involves drawing and redrawing of network number of times until the final network is obtained. A more rational approach may be considered as following:

(1) Preparing a work breakdown structure comprising the various compo-nents (activities) for the project,

(2) Preparing the complete list of activities required to complete the project.

(3) Establishing the technological relationships between the activities, which can be represented in a table consisting of two columns as shown in Table 1.1. All activities in the given sequence are listed in the second column and corresponding activities, on which they depend, are entered in the first column.

(4) Preparing a network representing all the activities and events in their logical sequence. Such a network is popularly known as Graphic Model of the Project, which is made of arrows, circles, dotted line arrows, hexagons, path, etc. These symbols denote their respective significance in the diagram (discussed later). This can be made as following:

   (i) The first activity of column-1 is connected to the activity opposite to it in the second column. The connected activity is then located in the first column and this activity is joined to the corresponding activity in the second column. This is repeated till one sequence is covered which gives one path.

Table 2

Column-1 Dependency	Column-2 Activities
–	A
–	B
–	C
A, B	D
B	E
B	F
A, B, C, F	G
B	H
E, H	I
E, H	J
C, D, F and J	K
K	L

(ii) The process is started with next activity from the top in the first column, which has not been connected in the first sequence, and the step (i) is followed and the sequence is completed.

(5) Step (ii) is repeated over and over again until all activities in the first column have been connected.

A simple illustration in the (Example 1) may be given as in Table 2.

A network diagram can be drawn as shown in Fig. 7.

Network Analysis

Following network computations are required for network analysis:

(1) The amount of time likely to be consumed by each activity.

(2) The earliest time by which an event can take place.

(3) The latest time by which an event can be completed without causing delay in scheduled date of completion.

(4) The amount of slack available between each event scheduled to be completed.

(5) Identification of critical and sub-critical paths through the network.

(6) The amount of total float available between various jobs.

(7) Systematic revision to meet the project objectives in terms of time-cost trade-off.

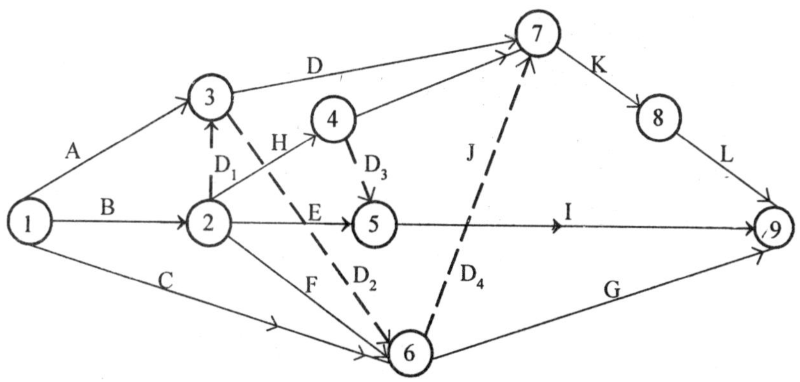

Fig. 7 : Network Diagram.

ESTIMATING TIME OF THE ACTIVITIES

After developing the network on the lines discussed above, the next logical step is to estimate the time required for completion of various activities comprising the project. The time assigned to activities should be realistic rather than desirable, *i.e.,* should be acceptable to those, responsible to carry out the jobs. Time of activities may be established either from past records or arrived from elemental data. In the absence of standard data, time for new jobs (where past records are not available) needs to be estimated. The time required for the performance of an activity under normal availability of resources, as planned, is called the "Activity Elapsed Time", t_e expressed in convenient units such as hours, days, weeks, months, etc., depending upon the size and nature of the project. It is recorded below the activity arrow.

The estimation of activity duration takes into account the following assumptions:

(i) Activity is sufficiently well defined.

(ii) The estimate of activity duration is independent of any influence from the preceding or succeeding activity.

(iii) Resources required to carry out the activity are available.

(iv) The estimate of activity duration includes only the normal delays and interruptions due to breakdowns, absenteeism, etc. The effects of unforeseen events such as fires, strikes, etc., is not considered.

(v) The time estimate is single time for the projects of repetitive nature, with the assumption that the project managers have a fair idea of the activity durations from their earlier experience.

(vi) For new activities, it may be difficult to establish one time estimate with reasonable accuracy and, thus, multiple, time estimates, based on PERT concept of Optimistic, Pessimistic, and most likely time should be used for such activities.

The activity time estimates are of great importance as they constitute the basis for the subsequent analysis on network. The successful completion of the project on schedule is very much dependent on this basic data, from which the activity time estimates are established.

The activity time estimates should be carried out by the persons most familiar with, or actually responsible for the performance of the activity, and/or who have a sound knowledge of the process involved in the completion of the project.

Time Uncertainty

New projects, not attempted earlier (such as Research and Development), are associated with an element of uncertainty of the time required for their accomplishment due to the absence of historic data or previous experience. This uncertainty for time estimates would result in a incorrect estimation of the duration of the project as a whole. Such an uncertainty factor can be dealt with a statistical method. When it is difficult to estimate the elapsed time for an activity precisely, its likelihood of achievement is expressed in three-time estimates rather than a positive assurance. These are optimistic time (a), pessimistic time (b) and most likely time (m).

1. Optimistic Performance Time (a)

It is minimum possible time an activity would take under ideal conditions, *i.e.,* if execution goes extremely well. Though, such a time is most unusual one, yet it is possible to happen (probability of 1 in 100 cases).

2. Pessimistic Time (b)

It is the largest time required by an activity on the assumption that every thing goes wrong (except in case of natural calamities). Such a time estimate again has the minimum probability like an optimistic performance time.

3. Most Likely Time (M)

It is the estimated time required by an activity under normal conditions. Even if the job is repeated under identical conditions, the time required would be almost the same.

The range, specified by the optimistic and pessimistic estimates (a and b, respectively), should essentially represent every possible estimate of the duration of the activity.

The most likely estimate m need not coincide with the midpoint $(a + b)/2$, and may occur to its left or right. Because of these properties, it is intuitively justified that the duration for each activity may follow a *beta* distribution with its unimodel point occurring at m and its end points at a and b.

Fig. 8 shows three cases of *beta* distribution, which are: (a) symmetric, (b) skewed to the right, and (c) skewed to the left.

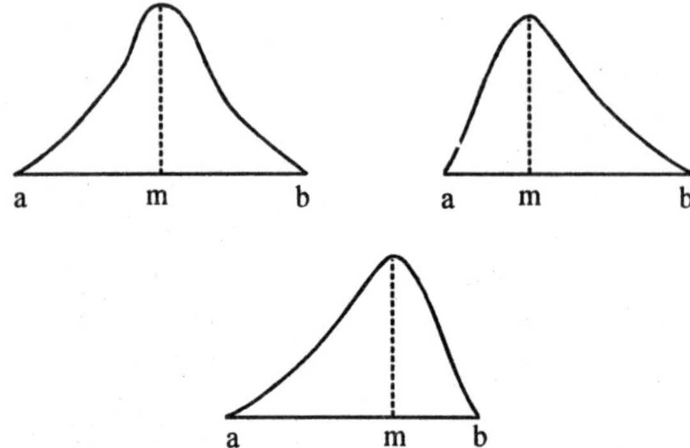

Fig. 8 : (a) Symmetric, (b) Skewed to Right, (c) Skewed to Left.

It has been observed that three time estimates statistically conform to skewed to right and expected time divides the area under curve into two equal parts, *i.e.,* there will be 50% chance for an activity to take longer than expected time. The expressions for expected time, t_e and, variance, *V*, can be developed as following: The mid-point (a + b)/2 is assumed to weigh half as the most likely point *m*. Thus expected time t_e represents the mean of the beta distribution and its value is more close to *m*. t_e is the value that is used in all the network calculations.

SOLVED EXAMPLES

Example 1:

Consider the network in Fig. 1. Time and cost (normal as well as crash) for the various activities are listed in Table 1 and the weekly incremental cost is worked out and given in last column of Table 1.

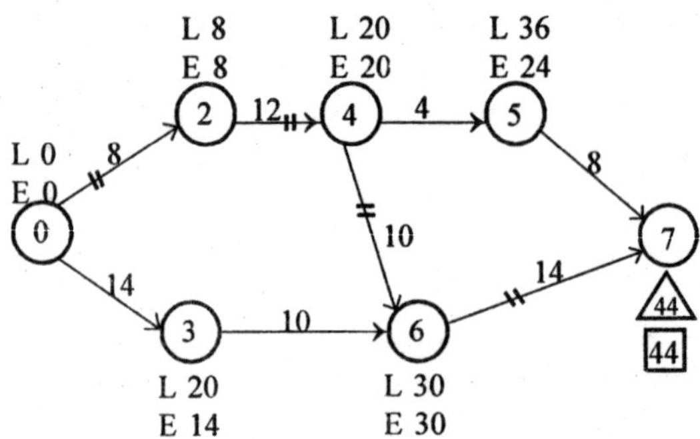

Fig 1 : (a) Diagram Showing Normal Time.

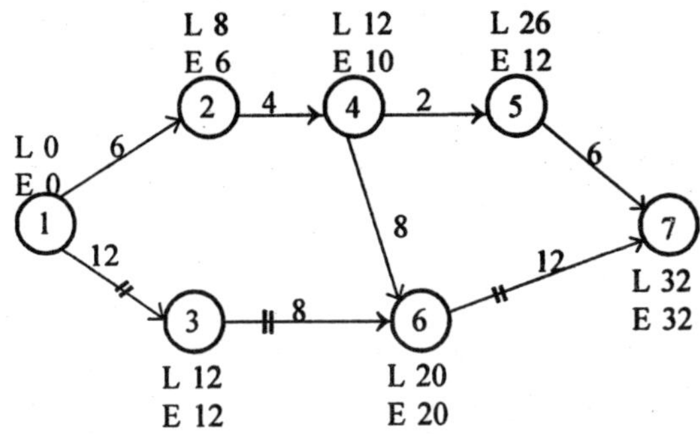

Fig 1 (b) : Diagram Showing Crash Times.

Solution:

The network in Fig. 1(a) shows the critical path calculations under normal conditions and the critical path is 1-2-4-6-7 and the project's earliest expected finish time is 44 weeks and its associated (normal) cost is 9200. Similarly, when all the activities are crashed, a new critical path 1-3-6-7 is obtained and it is likely to be completed in 32 weeks by incurring a total cost of 17,000. The next step is to find out whether or not the project's completion time can be reduced to 32 weeks without crashing all the activities on the network. If success is achieved in such an exercise, the project's completion time would be reduced as scheduled but without increasing the total project's cost to 17,000.

Table 1

Activity	Time (Weeks)		Cost		Weekly Incremental
	Normal	Crash	Normal	Crash	Cost (Slope)
1-2	8	6	1200	2000	400
1-3	14	12	1300	2800	750
2-4	12	4	1500	3900	300
3-6	10	8	1000	1900	450
4-5	4	2	900	1400	250
4-6	10	8	1100	2300	600
5-7	8	6	1800	2100	150
6-7	14	12	400	600	100
			9200	17,000	

In order to reduce the project time, attention is given to those critical activities, which are least costly ones to be crashed. In Fig. 1 (b), these are critical activities 1-3, 3-6, and 6-7 and the least expensive activity is 6-7 involving only a cost of 100 per week as compared to others, which will involve 750 and 450. However, activity 6-7 can be crashed subject to a maximum period of 2 weeks.

Accordingly, fresh network is prepared and Fig. 2 reveals that the earliest expected date of network ending comes down to 42 weeks from the original estimated network of 44 weeks with the increased cost of 9400 (9200 + 200). This network may give rise to a new critical path. However, in the present case, still it is the same, *i.e.,* 1-2-4-6-7.

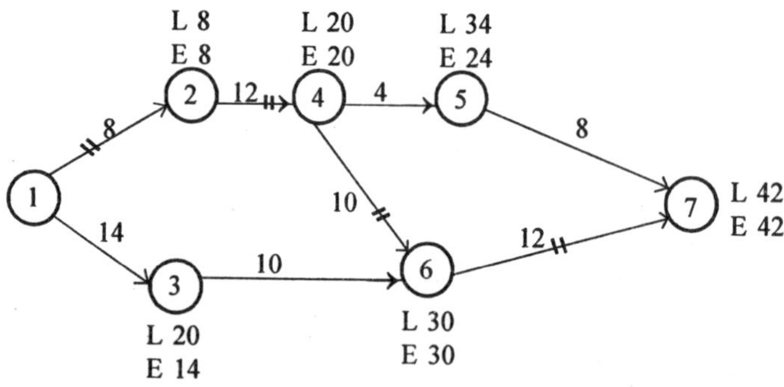

Fig. 2 : Next Expensive C-path 1-2-4-6-7 Cost 9200+200 = 9400. Activity 6-7 Rs. 200 for 2 Weeks.

The next least expensive activity to crash on this critical path is then observed from the computation table (Table 1). It is observed that activity 2-4 can be crashed for a period of 8 weeks and it would incur an additional cost of 2400 (300 × 8). Accordingly, the network is prepared (Fig. 3), which reveals that the earliest expected time of network ending event is reduced from 42 weeks to 36 weeks, but with a total project's cost of 11,800 (9400 + 2400).

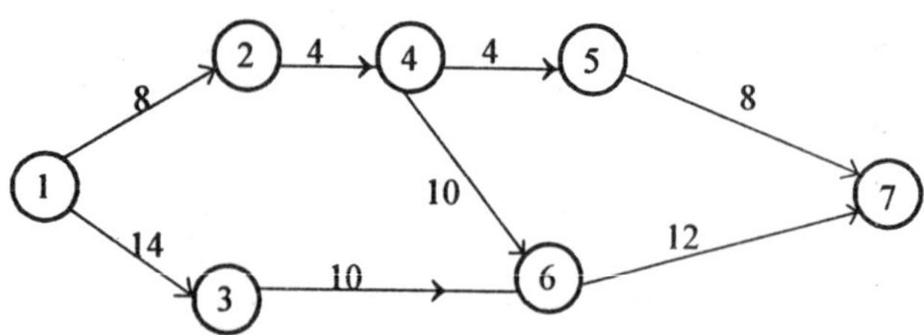

Fig 3 : Next Expensive Alternative Cost 9400 + 2400 = 11,800. Activity 2-4 Rs. 2400 for 8 Weeks.

While continuing this process, the next least expensive activity in the critical path 1-3-6-7 is 3-6 for a period of 2 weeks with an additional cost of 450 per week. The revised network (Fig. 4) brings out that the end event time has been further reduced to 34 weeks and, by this time, the cost outlay increases to 12, 700 (11,800 + 900). But our ultimate aim is to crash the network down to 32 weeks[1] time as proposed in the first instant.

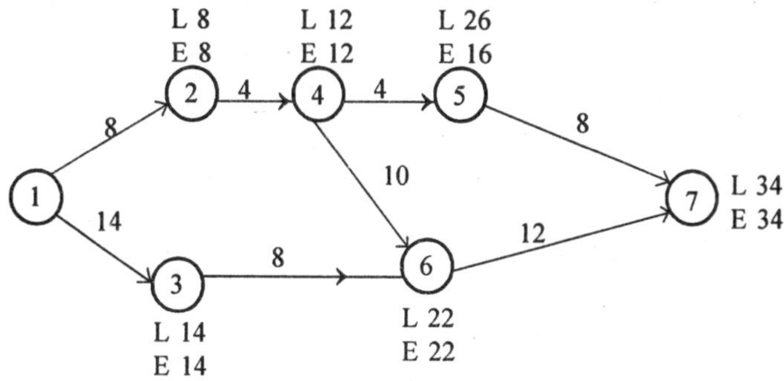

Fig 4 : Next Expensive Cost 11,800 + 900 = 12,700.
Activity 3-6 Rs. 900 for 2 Weeks.

By now the network has two critical paths, namely 1-2-4-6-7 and 1-3-6-7. Proceeding as before, while on the former critical path, activity 1-2 for each week at an additional cost of 400 is the least expensive one, on the latter, it is activity 1-3 for each week at an additional cost of 750. As such, the next network is prepared (Fig. 5) by crashing both the activities 1-2 and 1-3 so that the total crashed period comes down to 32 weeks, as originally envisaged. This way, the cost of project is further increased from 12,700 to 15,000 (12,700 + 2 × 400 + 2 × 750).

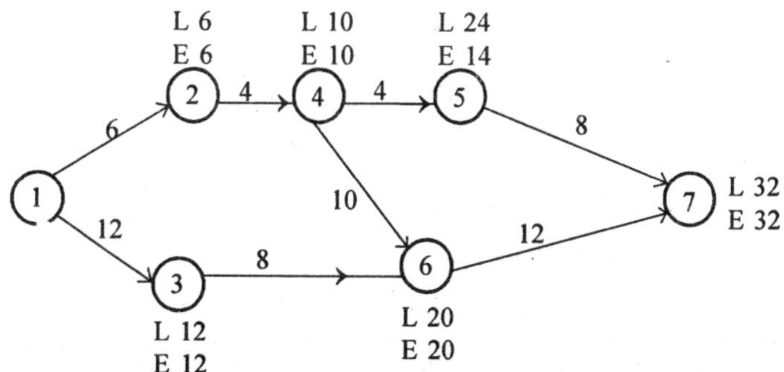

Fig 5 : Next Expensive Cost 12,700 + 2 × 400 + 2 × 750 = 15,000.
Activity 1-2 Rs. 800 for 2 Weeks; Activity 1-3 Rs. 1500 for 2 Weeks.

A summary of crashing all the activities is prepared as shown in Table 1.8.

Table 1.8

By crashing	6-7	for 2 weeks	= 200
By crashing	2-4	for 8 weeks	= 2400
By crashing	3-6	for 2 weeks	= 900
By crashing	1-2	for 2 weeks	= 800
By crashing	1-3	for 2 weeks	= 1500
		9200 + 5800 = 15,000	

It is evident from the data, the project duration can be crashed as conceived and have still saved an amount of 2,000 (17,000 – 15,000). If all the activities would have been crashed, this saving would not have been possible. Therefore, the utility of CPM lies in trading off cost and time relationship of given project.

However, in addition to direct cost, a project has to incur indirect and utility cost. All these costs, in fact, interact to arrive at the optimum time duration of a project. As such, the remaining two costs have also to be plotted on the graphs and final decision taken accordingly.

Example 2:

Maximize $Z = 12x_1 + 32x_2$

Subject to $6x_1 + 14x_2 \leq 65$

$5x_1 - 2x_2 \leq 20$

$x_1, x_2 \geq 0$

x_1, x_2 *are integers.*

Solution:

The optimal solution to the problem is $x_1 = 5$, $x_2 = 2½$ with Z = 140. However, if pure integer solution is required, then x_2 may be rounded off to either 2 or 3, the resulting solution with 3 becomes infeasible. The nearest pure-integer feasible solution to (5, 2½) is (4, 2) with the corresponding Z = 112. However, the correct optimal integer solution is (0, 4) with the value of objective function Z = 128.

This is a simple problem of IP and can be solved easily but in more complex problems, this is usually not possible. Thus, this indicates the requirement for special algorithms for solving Integer Programming Problem.

Example 3:

Consider the following Integer Linear Programming

Maximize $7x_1 + 5x_2$

Subject to

$$x_1 + x_2 \leq 7$$

$$2x_1 + 6x_2 \leq 61$$

$$x_1, x_2 \geq 0 \text{ and integer.}$$

Solution

The integral linear programming solution of the problem is the integral linear programming solution is shown by dots and the associated linear programming solution space (LPO) is defined by leaving the integer constraints. The optimum LPO solution from $x_1 = 4.75$, $x_2 = 2.25$ and Z = 44.5.

The Branch and Bound procedure is based on dealing with linear programming problem only. The optimum linear programming solution of the above (x_1 = 4.75, x_2 = 2.25 and Z = 44.5) does not satisfy the integer requirements and, thus, the branch and bounding algorithm involves "modifying" the linear programming solution space in a way to allow the identification of the integer linear programming optimum. It can be observed and verified that the objective function value of any integer solution to the problem has to be less than 44.5. The observation/ verification can be summarized as below:

1. In general, the integer restriction only leads to worsening of the objective function value. As such, for a maximization problem, the linear programming optimal solution is an upper bound, whereas for a minimization case, it gives the lower bound.
2. Next, the optimal solution of the linear programming as obtained above is examined. The non-integer variable is then selected (if only one is there). In case of more than one variables are non-integer, any one among them can be chosen. In the above example, x_1 and x_2 both are non-integers. For an integer solution, the value of x_1 greater than 4 and less than 5 is ruled out, and thus it can be specified that $x_1 \leq 4$ or $x_1 \geq 5$. These can be used to partition the solution space of the original linear programming into two distinct problems as follows:

   P-1

$$\text{Maximize } 7x_1 + 5x_2$$

Subject to

$$x_1 + x_2 \leq 7$$
$$10x_1 + 6x_2 \leq 61$$
$$x_1 \leq 4$$

x_1, x_2 are non-negative

P-2

$$\text{Maximize } 7x_1 + 5x_2$$

Subject to

$$x_1 + x_2 \leq 7$$
$$10x_1 + 6x_2 \leq 61$$
$$x_1 \leq 5$$

x_1, x_2 are non-negative

The above partitioning does not rule out any integer solution to the original problem. Now, if both the problems are solved, the various possible outcomes are listed below:

(i) Both the problems have feasible solution. Let the objective function values obtained for *P*–1 and *P*–2 be *Z*–1 and *Z*–2, respectively. However, it may be noted that both these values (of *Z*–1 and *Z*–2) have to be less than or equal to *Z*, as the objective function value can never improve due to an additional constraint. The solution, thus, may indicate any of the following three outcomes:

(a) Both *P*–1 and *P*–2 have integer solution.

(b) Only one of the problems gives an integer solution.

(c) None of the problems gives integer solution.

In case (a), max. (*Z*–1, *Z*–2) can be found out, and the optimal solution is the one showing the maximum.

In case of (b), the integer solution gives a feasible solution to the integer programming problem, say *P*–1 gives an integer solution. Thus, from *P*–1, no further branching is necessary, for branching from *P*–2, first the maximum objective function value arising from *P*–2 noted, say it is *Z*–2. If $Z\text{–}1 \geq Z - 2$, there is no need to explore *P*–2 further, as there cannot be any gain in doing so and the solution corresponding to *P*–1 is the required optimal solution. However, if $Z\text{–}1 < Z\text{–}2$, then *P*–2 has to be branched as the possibility of reaching a feasible integer solution with objective function value > *Z*–1 cannot be ruled out. Any one of the non-integer variables may be chosen, and branching can be carried out as shown earlier.

In case of (c), there is a need to branch from both the methods in exactly the same way as shown earlier. The new upper bounds for the branches from *P*–1 and *P*–2 are *Z*-1 and *Z*–2 respectively.

(ii) Both the problems are infeasible. As the whole solution space has been covered, it can be said that integer programming does not haver any feasible solution.

(iii) Only one of the problems is feasible. If the solution is integer, it would be the required one. When the solution is not integer, branching is done from the point with the upper new bound as the objective function value of the feasible problem (*P*–1 and *P*–2).

In the above example, the solution may be obtained as follows:

Step-1

Optimal solution as liner programming is determined as,

$$x_1 = 4.75,\ x_2 = 2.25, \text{ and } Z = 44.5$$

Step-2

Since the solution is not integer valued, step-2 is followed and x_1 is chosen, since $[x^*_1] = [19/4] = 4$ and the two problems are formed as,

$$LP_1\ x_1 \leq 4$$

$$LP_2\ x_1 \geq 5$$

LP_1 is solved and LP_2 can be represented graphically. It can observed that the two spaces contain the same integer feasible points of the integer linear programming model, *i.e.*, with respect to integer linear programming problem dealing with LP_1 and LP_2 is the same as dealing with the original linear programming space. The main difference is that the new bounding constraints ($x_1 \leq 4$ and $x_1 \geq 5$) will improve the chance of forcing the optimum extreme points of LP_1 and LP_2 toward satisfying the integer constraints. Further, the fact of bounding constraints being in the "immediate" vicinity of the continuous Linear programming space, optimum will improve their chances of giving "good" integer solution.

As the new restrictions $x_1 \leq 4$ and $x_1 \geq 5$ are mutually exclusive, LP_1 and LP_2 should be solved as two separate linear programming problems.

Step-3

LP_1 is solved and it is found that it has a feasible solution,

$$x_1 = 4, x_2 = 3 \text{ and } Z = 43 \qquad [A]$$

This solution satisfies the integer constraints and hence it is recorded at this step and $Z = 43$ is taken as lower bound.

Solution of LP_2 gives

$$x_1 = 5, x_2 = 11/6$$

and $\quad Z =$

Since this solution is not integer-valued, step-4 is followed.

Step-4

Now x_2 is selected. Since

$$[x^*_2] = [11/6] = 1$$

Thus the following problems are formed:

Maximize $7x_1 + 5x_2$

Subject to

$$x_1 + x_2 \leq 7$$

$$10x_1 + 6x_2 \leq 61$$

LP_3 $x_1 \geq 5, x_2 \leq 1$

LP_4 $x_1 \geq 5, x_2 \geq 2$

x_1, x_2 are non-negative

LP_3 is selected and the solution gives,

$x_1 = 5, x_2 = 1$ and $Z = 40$ [B]

This solution satisfies the integer conditions and thus it is recorded.

LP_4 has no feasible solution

Now, there is no more problem to be formed and thus the process ends.

At the time of ending process, it is observed that only two feasible integer solutions [A] and [B] have been noted. The "best one" of these two feasible solutions gives us the required optimum solution to the given integer programming problem.

Example 4:

A firm manufacturing bicycles has specialised in two types of bicycles: Simple model and delux model, both requiring two types of parts (raw materials). A simple model requires 12 and 14 units of first and the second type of raw materials, while the delux model requires 28 units and 14 units of first and second kind of raw materials. In a day, the manufacturer has a supply of 84 and 70 units of the two types of raw materials. Each unit of simple model contributes a profit of Rs. 200, while the delux model contributes Rs. 320. Find the optimum number of two types of bicycles to be produced in a day to maximize the total profit.

Solution:

Above problem in the form of integer Programming Problem can be formulated as following:

If x_1 = number of simple model bicycles

x_2 = number of deluxe model bicycles

Maximize $Z = 200x_1 + 320x_2$

Subject to

$$12x_1 + 28x_2 \leq 84$$

$$14x_1 + 14x_2 \leq 70$$

$x_1, x_2 \geq 0$ and are integers.

When computed as a Linear Programming Problem, the graphic approach gives a solution with $x_1 = 3.5$ and $x_2 = 1.5$ and the value of the objective function is Rs. 1180. However, as this solution is non-integer, it is not acceptable. If the values are rounded off to integer values, it will result in $x_1 = 3$ and $x_2 = 1$, rounding it off to next higher values will result into infeasible solution, and the value of the objective function is Rs. 920. However, this is not an optimal solution. In order to identify the integer optimum solution to the problem, moving from the non-integer optimum point (point B) towards the origin with slope of the isoprofit function, will give the optimum integer solution as the first integer point touched with the iso-profit function. This point is C, where $x_1 = 5$ and $x_2 = 0$ and the profit is 1000. It may be seen that the integer optimum solution yields a lower profit than a non-integer optimum solution. This cost is attached with the additional constraint of integer values to the decision variables.

Example 5:

Construct the network diagram having activities A, B, C, ... and L. such that the following relationships are satisfied.

(1) *A, B, and C the first activities can start concurrently.*

(2) *A and B precede D*

(3) *A and B precede G*

(4) *B precedes E, F and H*

(5) *F and C precede G*

(6) *E and H precede I and J*

(7) *C, D, F, and J precede K*

(8) *K precedes L*

(9) *I, G, and L are the terminal activities of the project.*

Solution:

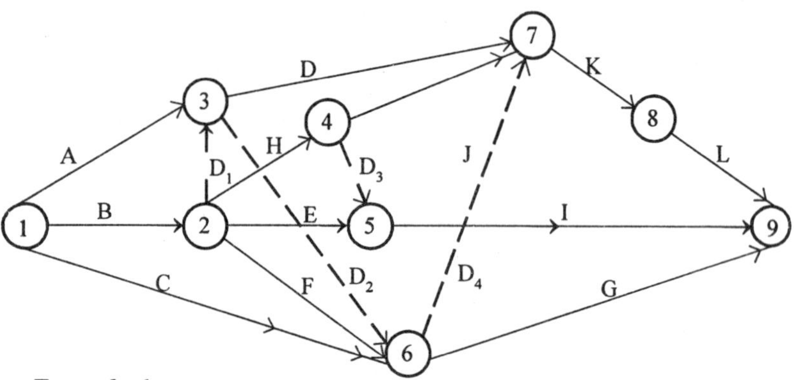

Example 6:

Fig. 6 : Represents the network of a small project:

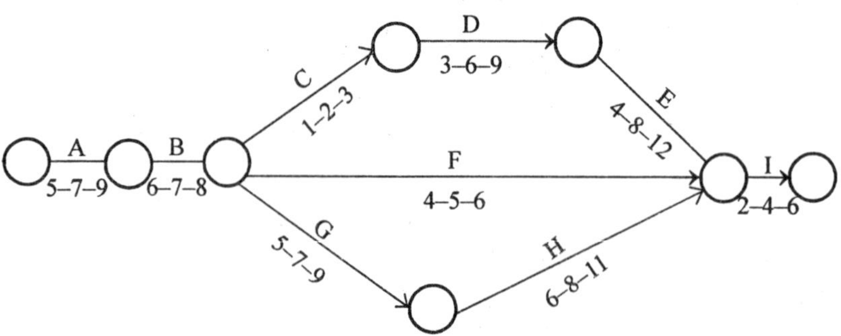

Fig. 6 : Network of a Small Project.

Solution:

Three time estimates—optimistic (a), most likely (m), and pessimistic (b) of each activity are incorporated into the network itself. Calculate expected time (t_e) for each activity. t_e of all activities of network can be tabulated as given in Table-1.

Table 2 : t_e Values of Network

Activity	a m b	Expected time (t_e)
A	5 - 7 - 9	7
B	6 - 7 - 8	7
C	1 - 2 - 3	2
D	3 - 6 - 9	6
E	4 - 8 - 12	8
F	4 - 5 - 6	5
G	5 - 7 - 9	7
H	6 - 8 - 10	8
I	2 - 4 - 6	4

t_e of each activity can now be entered into network as shown in Fig. 7.

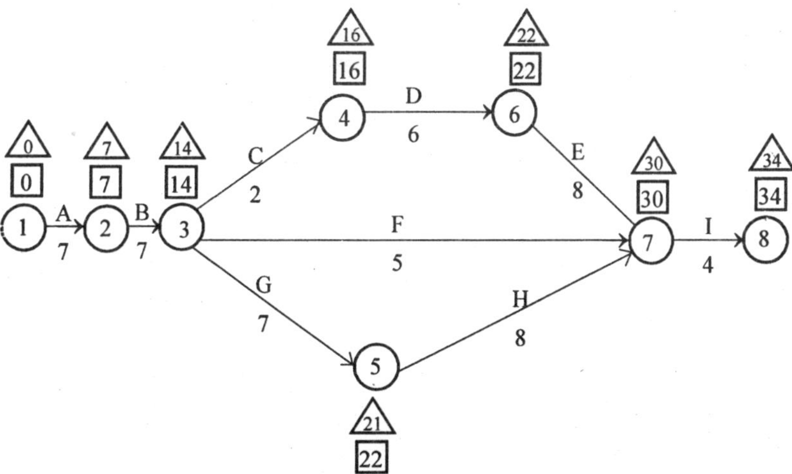

Fig 7 : Illustration of Network with Three Time Estimates Converted into Single Expected Time.

Calculation for earliest starting date (*ES*), Earliest finishing date (EF_{ij}), Latest finishing date (LF_{ij}), Latest starting date (LS_{ij}), total float, and independent float for the network shown in Fig. 1.13 are given in Table 1.5.

Table 1.5

Activity	Activity duration D_{ij}	Earliest Start ES_i=TE of the tail event	Earliest Compl. EC_{ij} = $ES_i + D_{ij}$	Lastest Start LS_{ij}	Lastest Compl. LC_j	Total float TF_{ij}	Free float FF_{ij}	Independent FI
1-2	7	0	7	0	7	0*	0	0
2-3	7	7	14	7	14	0*	0	0
3-4	2	14	16	14	16	0*	0	0
3-5	7	14	21	13	22	1	0	0
4-6	6	16	22	16	22	0*	0	0
3-7	5	14	19	25	30	11	11	11
5-7	8	14	29	22	30	1	1	0
6-7	8	20	30	22	30	0*	0	0
7-8	4	30	34	30	34	0*	0	0

*Critical Activity

The critical activity must have a zero total float. The free float must also be zero when the total float is zero. However, the converse is not true, *i.e.*, a non-critical activity may have zero free float (in above Table 1.5, non-critical activity 3-5 has zero free float).

Example 7(a):

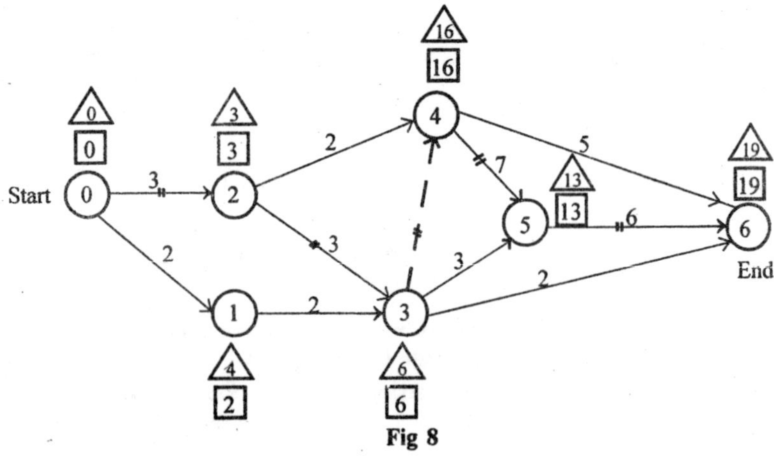

Fig 8

Solution:

Table 3 gives the critical path calculations together with the floats for the non-critical activities. Columns 1, 2, 3, and 6 are obtained from the network calcuations. The remaining information can be determined from the foregone formulas.

Table 3

		Earliest		Latest			
Activity (i, j) (1)	**Duration D_{ij} (2)**	**Start □ES_i (3)**	**Completion EC_{ij} (4)**	**Start LS_{ij} (5)**	**Completion ΔLC_j (6)**	**Total Float TF_{ij} (7)**	**Free Float FF_{ij} (8)**
(0.1)	2	0	2	2	4	2	0
(0.2)	3	0	3	0	3	0*	0
(1.3)	2	2	4	4	6	2	2
(2.3)	3	3	6	3	6	0*	0
(2.4)	2	3	5	4	6	1	1
(3.4)	0	6	6	6	6	0*	0
(3.5)	3	6	9	10	13	4	4
(3.6)	2	6	8	17	19	11	11
(4.5)	7	6	13	6	13	0*	0
(4.6)	5	6	11	14	19	8	8
(5.6)	6	13	19	13	19	0*	0

*Critical Activity

For constructing the time chart, the first step is to consider the scheduling of the critical activities. Next, the non-critical activities are considered by indicating their *ES* and *EC* time limits on the chart. The critical activities are shown with solid lines(→). The time ranges for the non-critical activities are shown by simple lines, indicating that such activities may be scheduled within those ranges provided that the precedence relationships are not disturbed.

The dummy activity (3-4) consumes no time and hence is shown by a vertical line. The numbers shown with the non-critical activities represent their durations.

The role of the *total* and *free* floats, in scheduling non-critical activities is explained in terms of two *general* rules:

(i) If the total float equals to the free float, the non-critical activity can be scheduled anywhere between the earliest start and latest completion times.

(ii) If the free float is < total float, the starting of the non-critical activity can be delayed relative to its earliest start time by no more than the amount of its free float without affecting the scheduling of its immediately succeeding activities.

In this example, rule-2 applies to activity (0, 1) only, whereas all others are scheduled according to rule-1. The reason is that activity (0, 1) has zero free float. Thus, if the starting time for (0, 1) is not delayed beyond earliest time ($t = 0$), the immediately succeeding activity (1, 3) can be sheduled anywhere between its earliest start time ($t = 2$) and latest completion time ($t = 6$). On the other hand, if the starting time of (0, 1) is delayed beyond $t = 0$, the earliest start time of (1.3) must be delayed relative to its earliest start time by atleast the same amount. For example, if (0, 1) starts at $t = 3$, it terminates at $t = 3$, and (1, 3) can then be scheduled anywhere between $t = 3$ and $t = 6$. This type of restriction does not apply to any of the remaining non-critical activities because they all have equal total and free floats.

This can also be observed, since (0, 1) and (0, 2) are the only two per-missible activities whose permissible time spans over lap.

In fact, the free float, being less than total float, gives a warning that the scheduling of the activity should not be finalized without first checking its effect on the start times of the immediately succeeding activity. This valuable information can be secured only through the use of critical path computations.

EXERCISES

1. A nationalised bank wishes to plan and schedule the development and installation of a new computerized cheque processing system. The changeover in cheque-processing procedures requires employment of additional personnel to operate the new system, development of new systems (computer software), and modification of existing cheque sorting equipment.

   The activities required to complete the project along with three time estimates and the precedence relationship among the activities have been determined by bank management and are given in the following table:

Activity	*Description*	*Predecessor*	*Time Estimates (days)*		
			Optimistic	*Most Likely*	*Pessimistic*
A	Position recruiting	–	5	8	17
B	System development	–	3	12	15
C	System training	A	4	7	10
D	Equipment training	A	5	8	23
E	Manual system test	B, C	1	1	1
F	Preliminary system changeover	B, C	1	4	13
G	Computer-personnel interface	D, E	3	6	9
H	Equipment modification	D, E	1	2.5	7
I	Equipment testing	H	1	1	1
J	System debugging and installation	F, G	2	2	2
K	Equipment changeover	G, I	5	8	11

(a) Draw a network diagram for this project and find the critical path and its length.

(b) Calculate the total and free floats for non-critical activities.

(c) What is the probability that the length of the critical path does not exceed 40 days?

2. How does PERT technique help a business manager in decision-making?

3. Critically comment on the assumptions on which PERT/CPM analysis is done for projects.

4. PERT now has proven to be an effective management tool that can be utilized by companies of all sizes in almost every industry. Discuss.

5. 'PERT provides the framework with which a project can be described, scheduled and then controlled?' Discuss.

6. A promoter is organizing a sports meeting. The relationship among the activities and time estimates in days are shown below in the table:

Activity	*Description*	*Immediate Predecessor*	*Activity Time (days)* Optimistic	Likely	Pessimistic
A	Prepare draft programme	–	3	7	11
B	Send to sports organizations and wait for comments	A	14	21	28
C	Obtain promoters	A	11	14	17
D	Prepare and sign documents for stadium hire	A, C	2	2	2
E	Redraft programme and request entries	B	2	7/2	8
F	Enlist officials	D, E	10	14	21
G	Arrange accommodations for touring teams	E	3	4	5
H	Prepare detailed programme	E, F	4	9/2	8
I	Make last-minute arrangements	G, H	1	2	4

(a) Find the expected activity durations and their variances.

(b) Draw a PERT network diagram and find the critical path. What is the expected length of the critical path and what is its variance?

(c) What is the probability that the length of the critical path does not exceed 56 days?

8. What are the basic assumptions underlying the "expected time" estimate?

9. Discuss the role of statistical technique in PERT.

10. Explain the following in the context of project management: (i) Activity variance, (ii) Project variance.

11. 'PERT' takes care of uncertain durations". How far is this statement correct? Explain with reasons.

12. How does PERT provides for uncertainty in activity time estimates? What is the rationale for using beta probability distribution?

13. A management student identifies following list of activities and sequen-cing requirements along with the time estimates for various activities related to the completion of his project work.

Activity	*Description*	*Immediate Predecessor*	*Activity Time (days)* *Optimistic*	*Likely*	*Pessimistic*
A	Search of list	–	3	6	9
B	Dedicating the project	–	2	4	12
C	Preliminary work	B	1	1.5	5
D	Formal proposal	C	1	2	3
E	Project committee's approval	A, D	1.5	2	4.5
F	Progress report	E	0.5	1	1.5
G	Format research	A, D	4.5	5	11.5
H	Data collection	E	2	5	8
I	Analysis	G, H	4	5.5	10
J	Conclusion	I	1.5	2.5	4.5
K	Draft	I, F	2	3.5	8
L	Final draft	J, K	2.5	3	1.5
M	Presentation	L	0.5	1	1.5

With the help of an arrow diagram, determine the minimum time required to complete the project. From the network, identify the activities which can be delayed without affecting the project duration and the extent of delay that is possible for such activities.

14. A reactor and storage tank are interconnected by a 3" insulated process line that needs periodic replacement. There are valves along the lines and at the terminals and these need replacing as well. No pipe and valves are in stock. Accurate, as built, drawings exist and are available.

    This line is overhead and requires scaffolding. Pipe sections can be shop-fabricated at the plant. Adequate craft labour is available. You are the maintenance and construction superintendent responsible for his project. The works engineer has requested your plan and schedule for a review with the operating supervision.

    The plant methods and standards section has furnished the following, data. The precedent for each activity have been determined from a familiarity with similar projects.

Activity	*Description*	*Time (hrs.)*	*Precedents*
A	Develop required material list	8	–
B	Procure pipe	200	A
C	Erect scaffold	12	–
D	Remove scaffold	4	H, M
E	Deactivate line	8	–
F	Prefabricate sections	40	B
G	Place new pipes	32	F, K
H	Fit up pipe and valves	8	G, J
I	Procure valves	225	A
J	Place valves	8	I, K
K	Remove old pipe and valves	35	C, E
L	Insulate	24	G, J
M	Pressure test	6	H
N	Clean-up and start-up	4	D, M

(a) Sketch the arrow diagram of this project plan.

(b) Make the forward pass and backward pass calculations on this network, and indicate the critical path and its length.

(c) Calculate total float and free float for each of the non-critical activities.

15. Explain the following terms in PERT/CPM:

    (i) Earliest time,

    (ii) Latest time,

    (iii) Total activity time,

    (iv) Event slack, and

    (v) Critical path.

16. (a) What is float? What are the different types of floats?

    (b) Discuss in brief: (i) total float, and (ii) free float and explain their uses in network.

17. What is meant by the term critical activities, and why is it necessary to know about them?

18. Explain the significance of working out of float' in the network of project activities. Discuss in brief the different types of floats.

19. Explain the following terms in the context for project management:
    (i) Resources float,
    (ii) Activity variance,
    (iii) Project variance.
20. Briefly mention the areas of application of network techniques.
21. Compare and contrast CPM and PERT. Under what conditions would you recommend scheduling by PERT? Justify your answer with reasons.
22. What are the major limitations of the PERT model? Discuss.
23. Explain the following terms in PERT:
    (i) Three time estimates, (ii) Expected time, (iii) Activity variance.
24. A project has the following activities and other characteristics:

Activity	*Preceding Activity*	*Time Estimates (Weeks)*		
		Optimistic	*Most Likely*	*Pessimistic*
A	–	4	7	16
B	–	1	5	15
C	A	6	12	30
D	A	2	5	8
E	C	5	11	17
F	D	3	6	15
G	B	3	9	27
H	E, F	1	4	7
I	G	4	19	28

(a) Draw the PERT network diagram.
(b) Identify the critical path.
(c) Prepare the activity schedule for the project.
(d) Determine the mean project completion time.
(e) Find the probability that the project is completed in 36 weeks.
(f) If the project manager wishes to be 99 per cent sure that the project is completed on June 30, 1991, when should he start the project work?

25. Discuss various steps involved in the applications of PERT and CPM.
26. Highlight the difficulties encountered in using network techniques.
27. A small project consists of seven activities, the details of which are given below:

Activity	*Most Likely*	*Duration (days) Optimistic*	*Pessimistic*	*Immediate Predecessor*
A	3	1	7	–
B	6	2	14	A
C	3	3	3	A
D	10	4	22	B, C
E	7	3	15	B
F	5	2	14	D, E
G	4	4	4	D

(a) Draw the network, number the nodes, find the critical path, the expected project completion time and the next most critical path.

(b) What project duration will have 95 per cent confidence of completion?

28. The owner of a chain of fast food restaurants is considering a new computer system for accounting and inventory control. A computer company sent the following information about the computer system installation:

Activity	*Description*	*Immediate Predecessor*	*Optimistic*	*Times (days) Most Likely*	*Pessimistic*
A	Select the computer model	–	4	6	8
B	Design input/output system	A	5	7	15
C	Design monitoring systems	A	4	8	12
D	Assemble computer hardware	B	15	20	25
E	Develop the main programmes	B	10	18	26
F	Develop input/output routines	C	8	9	16
G	Create database	E	4	8	12
H	Install the system	D, F	1	2	3
I	Test and implement	G, H	6	7	8

(a) Construct PERT network diagram for this problem.

(b) Determine the critical path and compute the expected completion time.

(c) Determine the probability of completing the project in 55 days.

29. Draw network diagrams from the following list of activities:

Activity	*Predecessor Activity*		
	Set 1	*Set 2*	*Set 3*
A	–	–	–
B	–	–	–
C	–	–	–
D	A	A	A
E	B	A, B	A, B
F	B, C	A, B, C	B, C
G	D, E, F	D, E, F	•C
H	E, F	F	D, E.F

30. A company manufacturing plant and equipment for chemical processing is in the process of quoting a tender called by a public sector undertaking. Delivery data once promised is crucial and penalty clause is applicable. Project manager has listed down the activities in the project as under:

Activity	*Immediate Predecessor*	*Activity Time (Weeks)*		
		Optimistic	*Most Likely*	*Pessimistic*
A	–	1	3	5
B	–	2	4	6
C	A	3	5	7
D	A	5	6	7
E	C	5	7	9
F	D	6	8	10
G	B	7	9	11
H	E, F, G	2	3	4

Using PERT: (a) find out the delivery week from the date of commencement of the project, and (b) total float and free float for each of the non-critical activities.

31. Explain reasons for incorporating dummy activities in a network diagram. In what way do these differ from the normal activities?

32. State the circumstances where CPM is a better technique of project management than PERT.

33. Explain the basic logic of arrow networks.

34. Consider a project having the following activities and their time estimates:

Activity	*Immediate Predecessor*	*Activity Time (Weeks)* Optimistic	Most Likely	Pessimistic
A	–	3	4	5
B	–	4	8	10
C	B	5	6	8
D	A, C	9	15	10
E	B	4	6	8
F	D, E	3	4	5
G	D, E	5	6	8
H	D, E	1	3	4
I	G	2	4	5
J	F, I	7	8	10
K	G	4	5	6
L	H	8	9	13
M	J, K, L	6	7	8

(a) Draw PERT network diagram for the project.

(b) Compute the expected project completion time.

(c) What should be the due date to have 0.90 probability of completion?

(d) Find the total float and free float for all non-critical activities.

35. A sociologist plans a questionnaire survey consisting of the following tasks:

Activity	*Description*	*Immediate Predecessor*	*Likely*	*Duration (days) Minimum*	*Pessimistic*
A	Design of questionnaire	–	5	4	6
B	Sampling design	–	12	8	16
C	Testing of questionnaire and refinements	A	5	4	12
D	Recruiting for interviewers	B	3	1	5
E	Training of interviewers	D, A	2	2	2
F	Allocation of areas to interviewers	B	5	4	6
G	Conducting interviews	C, E, F	14	10	18
H	Evaluation of results	G	20	18	34

(a) For this PERT network find the expected task durations and the variances of task durations.

(b) Draw a network for this project and find the critical path. What is the expected length of the critical path? What is the variance of the length of the critical path?

(c) What is the probability that the length of the critical path does not exceed 60 days?

36. The activities of a project are tabulated below with immediate predecessors and normal and crash time cost.

Activity	*Immediate Predecessor*	*Normal*		*Crash*	
		Cost (Rs.)	*Time (days)*	*Cost (Rs.)*	*Time (days)*
A	–	200	3	400	2
B	–	250	8	700	5
C	–	320	5	380	4
D	A	410	0	800	4
E	C	600	2	670	1
F	B, E	400	6	950	1
H	B, E	550	12	1,000	6
G	D	300	11	400	9

(a) Draw the network corresponding to normal time.

(b) Determine the critical path and the normal duration and cost of the project.

(c) Suitably crash the activities so that the normal duration may be reduced by 3 days at minimum cost. Also find the project cost for this shortened duration if the indirect cost per day is Rs. 25.

37. A publisher is preparing to produce the second edition of a textbook. The activities required and their estimated time are as follows:

Activity	*Description*	*Immediate Predecessor*	*Activity Time (days)* *Optimistic*	*Likely*	*Pessimistic*
A	Assess market	–	1	3	2
B	Get reviews from users	A	1	2	1.5
C	Revamp old material and add new material	B	3	9	5
D	Obtain review and prepare final draft	C	4	12	6
E	Revise and expand problems	B	2	7	4
F	Copy edit final draft	D	1	2.5	1.5
G	Copy edit problems	E	0.5	1.5	1
H	Set type, proof and print book	F, G	5	9	6
I	Prepare instructor's manual	E	2	4	3
J	Produce instructor's manual	I	1	2	1.5
K	Complete book and instructor's manual	H, J	0	0	0

(a) Draw a PERT network diagram and determine the critical path.

(b) What is probability that the project will be completed within 21 months? 24 months? 27 months?

38. A medical association prepares an annual programme each year giving the monthly meeting dates, background on the speakers, an

abstract of their talks, and an alphabetic listing, both by name and medical college/hospital affiliation, of all dues paying members. The programme is mailed to these members as well as to selected individuals and organizations. The activities to be performed are listed below:

Activity	*Description* (*weeks*)	*Preceding Activities*	*Estimated Time*
A	Decide on general orientation for this year's programme	–	1
B	Get commitments from speakers and abstracts of their talks	A	4
C	Solicit advertising to appear in the programme	A	3
D	Mail out dues notices and wait for response	–	6
E	Prepare list of dues-paying members	D	1
F	Get copy to printer and proofread	B, C, E	2
G	Get programme printed and assembled	F	1
H	Prepare final mailing list	E	1
I	Staff envelopes and mail programmes	G, H	1

(a) Develop a PERT diagram to show the relationship between all activities. Specify the activities on the critical path and the project completion time.

(b) The programme chairman now claim that it will take him 6 weeks to get commitments and abstracts from the speakers. Will this delay the project completion time? How will it affect the critical path?

39. A promoter is organizing a sports meeting. The relationship among the activities and time estimates in days are shown below in the table:

Activity	*Description*	*Predecessor*	*Activity Time (days)* *Optimistic*	*Likely*	*Pessimistic*
A	Prepare draft programme	–	3	7	11
B	Send to sports organizations and wait for comments	A	14	21	28
C	Obtain promoters	A	11	14	17

D	Prepare and sign documents for stadium hire	A, C	2	2	2
E	Redraft programme and request entries	B	2	3.5	8
F	Enlist officials	D, E	10	14	21
G	Arrange accommodation for touring teams	E	3	4	5
H	Prepare detailed programme	E, F	4	4.5	8
I	Make last-minute arrangements	G, H	I	2	4

(a) Draw the network diagram for the project and compute expected completion time of the project.

(b) What should be the due date to have 0.90 probability of project completion?

(c) Find the total float and free float for all non-critical activities.

(d) What is the probability that the length of the critical path does not exceed 56 days?

40. A management student identifies the following list of activities and sequencing requirements along with the time estimates for various activities related to the completion of his project:

Activity	*Description*	*Predecessor*	*Activity Time (days)*		
			Optimistic	*Likely*	*Pessimistic*
A	Search of Literature	–	3	6	9
B	Deciding the project	–	2	4	12
C	Preliminary work	B	1	1.5	5
D	Formal proposal	C	1	2	3
E	Project committee's approval	A, D	1.5	2	4.5
F	Progress report	E	0.5	1	1.5
G	Formal research	A, D	4.5	5	11.5
H	Data collection	E	2	5	8
I	Analysis	G, H	4	5.5	10
J	Conclusion	I	1.5	2.5	4.5
K	Draft	I, F	2	3.4	8
L	Final Draft	J, K	2.5	3	1.4
M	Presentation	L	0.5	1	1.5

With the help of an arrow diagram, determine the minimum time re-quired to complete the project. From the network, identify the activities which can be delayed without affecting the project duration and the extent of delay that is possible for such activities.

41. A computer software company has broken down the process of integrating a computer system into its operation into several steps. Some of the steps cannot begin until others are completed, and these relationships are shown in the accompanying table. In addition, estimates of the most likely, optimistic, and pessimistic times required for each are listed below:

Activity	*Immediate*	*Expected Time (days)*		
		Optimistic	*Most Likely*	*Pessimistic*
A	–	1	3	4
B	–	3	4	11
C	A	2	5	8
D	A	1.5	3.5	8.5
E	B	5	7	9
F	B	2	5.5	6
G	C	1.5	2.5	6.5
H	C, D	3	4	II
I	G	4	6	8
J	H, E	3	4.5	9
K	F	5	6	7
L	I, J, K	1	3	11

(a) Draw a PERT chart for this project, and calculate the critical path

(b) By how much time activity F be delayed without delaying the project as a whole?

(c) If labour costs Rs. 1500 per week find the probability that the labour costs for this project will exceed Rs. 36,000.

(d) Company wishes to budget an amount for labour costs that will be sufficient with 95 per cent probability. How much should they budget?

42. The required data for a small project consisting of different activities are given below:

Activity	*Predecessor Activities*	*Normal*		*Crash*	
		Duration (days)	*Cost (Rs.)*	*Duration (days)*	*Cost (Rs.)*
A	–	6	300	5	400
B	–	8	400	6	600
C	A	7	400	5	600
D	B	12	1,000	4	1,400
E	C	8	800	8	800
F	B	7	400	6	500
G	D, E	5	1,000	3	1,400
H	F	8	500	5	700

(a) Draw the network and find out the normal project length and minimum project length.

(b) If the project is to be completed in 21 days with minimum crash cost which activities should be crashed to how many days?

43. A multinational FMCG company wishes to launch a new Fruit Yogurt in the coming season. A brief description of the activities associated with this project, their expected durations (in weeks) and their immediate predecessor(s) are given in the following table:

Activity	*Description (Weeks)*	*Predecessor*	*Expected Time*		
			Optimistic	*Likely*	*Pessimistic*
A	Management approval	–	2	2.5	4
B	Product concept test	A	3	4.7	5
C	Technical feasibility	A	2	2	3
D	Recipe finalization	C, B	1	1	2
E	Shelf life trials	D	8	12	15
F	Brand positioning study	5	4	5	7
G	Packaging key lines	F	11	2.5	
H	Agency advertisement development	F, E	4	8	10
I	Agency: layouts artworks	G	2	3	5
J	Advertisement test research	H	3	4	5

K	Cost finalization	D	1	1	2
L	Pricing decision	J, F	1	1.5	3
M	Marketing mix finalization	L	2	2.5	3
N	POS development	M	2	4	5
O	Launch plans	M	1	1	1.5
P	Branch communication	0	0.5	0.5	1
Q	Supplier's delivery of packaging	H	4	5	8
R	Production trail	D, Q	1.5	2	3
S	Management final approval	R	0.5	1	1.5
T	Final production	S	1	2.5	3
U	Stock movement	T	0.5	1.5	2
V	Position movement	T	0.5	1	1.5
W	Launch	U, V	1	2	4

Management of the company desires to know the realistic completion time for this project and detailed analysis of float times (if any).

44. Following is the list of various activities involved in the launch of a new Credit Card service by a company, their immediate predecessors and their expected durations (in days):

Activity	*Description (weeks)*	*Immediate*	*Activity*	*Time*	
	Predecessor	*Optimistic*	*Likely*	*Pessimistic*	
A	Conduct market research to determine size, potential and competition in the market	–	10	12	14
B	Define marketing strategy in terms of positioning and product features	A	14	15	17
C	Estimate expected volumes	B	2	3	4
D	Estimate additional manpower required	C	4	6	8
E	Identify modifications required to the present computer system based on expected volumes	C	10	12	14
F	Implement computer system modifications	E	20	25	27

G	Update physical facilities	C	10	17	20
H	Prepare instruction manuals for modified system	F	5	6'	7
I	Hire additional sales force and operations staff	D	7	12	14
J	Train sales force and operations staff	H, I	14	17	20
K	Advertise for suppliers	C	1	2	3
L	Check supplier samples	K	10	15	20
M	Identify key suppliers and define quality standards	L	3	5	7
N	Roll-out basic product to a few employees as a method of pre-testing	M, J	13	15	17
O	Generate and assess feedback	N	20	21	22
P	Modify and make changes wherever necessary	O	7	9	14
Q	Update procedures manual	P	2	3	4
R	Retain commercial staff	Q	2	2	2
S	Design communications to prospective card-holders	P	7	10	13
T	Design advertising schedule	s	5	7	9
U	Roll out final product-Launch	T, R, G	4	8	12

(a) Draw an arrow diagram for the project.

(b) Find the expected project completion time of the project.

(c) Determine the probability of completing the project in 165 days.

45. The time and cost estimates and precedence relationship of the different activities constituting a project are given below:

Activity	*Predecessor Activities*	*Time (in weeks)*		*Cost (in Rs.)*	
		Normal	*Crash*	*Normal*	*Crash*
A	–	3	2	8,000	19,000
B	–	8	6	600	1,000
C	B	6	4	10,000	12,000
D	B	5	2	4,000	10,000

E	A	13	10	3,000	9,000
F	A	4	4	15,000	15,000
G	F	2	1	1,200	1,400
H	C, E, G	6	4	3,500	4,500
I	F	2	1	7,000	8,000

(a) Draw a project network diagram and find the critical path.

(b) If a dead line of 17 weeks is imposed for completion of the project, what activities will be crashed, what would be the additional cost and what would be critical activities of the crashed network after crashing?

46. Following is the list of activities associated with the assembly of a space module for an upcoming mission. Draw the network diagram for the project.

Activity	*Description*	*Predecessor Activity*
A	Construct shell of module	–
B	Order life support system and scientific experimentation package from supplier	–
C	Order components of control and navigation system	–
D	Wire module	A
E	Assemble control and navigational system	C
F	Preliminary test of life support system	B
G	Install life support system in module	D, F
H	Install scientific experimentation package in module	D, F
I	Preliminary test of control and navigational system	E, F
J	Install control and navigational system in module	H, I
K	Final testing and debugging	G, J

47. An architect has been awarded a contract to prepare plans for an urban renewal project. The job consists of the following activities and their estimated times:

Activity	*Description*	*Immediate Predecessors*	*Time (days)*
A	Prepare preliminary sketches	–	2
B	Outline specifications	–	1
C	Prepare drawings	A	3
D	Write specifications	A, B	2
E	Run off prints	C, D	1
F	Have specification	B, D	3
G	Assemble bid packages	E, F	1

(a) Draw an arrow diagram for this project.

(b) Indicate the critical path, and calculate the total float and free float for each activity.

48. Listed in the table are the activities and sequencing requirement necessary for light system maintenance project at a particular stadium.

Activity	*Description*	*Predecessor Activity*
A	Assemble crew	–
B	Test lights for burned bulbs	–
C	Obtain needed bulbs	B
D	Paint light standards below banks	A
E	Replace burned bulbs	C
F	Deactivate system	B
G	Check all wiring for wear	A, F
H	Obtain needed wire	G
I	Clean lenses on lights	A, F
J	Remove worn wire	G
K	Cut new wire to needed lengths	J, H
L	Check insulators that support wires	J
M	Replace worn insulators	L
N	Replace old wire	M, K
O	Splice new wire with old	N
P	Insulate splices	O
Q	Paint light banks	P
R	Replace broken lenses	E
S	Reactivate system	Q, D, I, R
T	Clean up	R

Draw the arrow diagram of activities of the project.

49. Delhi Medical Association is considering to hold a conference. The following table gives the list of activities involved, their immediate predecessors and their duration (in days):

Activity	*Description*	*Predecessors*	*Duration (days)*
A	Design conference meetings and theme	–	3
B	Design front cover of the conference proceedings	A	2
C	Prepare brochure and send request for papers	A	6
D	Compile list of distinguished speakers/guests	A	3
E	Finalize brochure and print it	C, D	7
F	Make travel arrangements for speakers/guests	D	4
G	Despatch brochures	E	3
H	Receive papers for conference	G	25
I	Edit papers and assemble proceedings	F, H	10
J	Print proceedings	B, I	20

(a) Prepare a network diagram showing the inter-relationships of the various activities.

(b) Find the total time required to hold the conference.

(c) Compute the total float for the non-critical activities.

50. Draw an arrow diagram for the following project related to the introduction of a new product.

Activity	*Description*	*Predecessor Activity*
A	Develop plan for introduction of the project	–
B	Prepare product drawings	A
C	Test and approve packages	B
D	Approve unit costs	C
E	Prepare and send questionnaires in the field	D
F	Evaluate questionnaires	E
G	Finalise labels, shipping boxes and cartons	F
H	Plan media advertising schedule and sales literature	F
I	Assess manpower needs, and hire and train employees	F

J	Design and develop production equipment	B
K	Arrange and instal production equipment	J
L	Debug equipment	K
M	Design and develop product and quality standards	J
N	Manufacture labels, shipping boxes and cartons	G
O	Order, receive and check raw materials	I, L
P	Prepare and check warehouse space	F
Q	Conduct trial run	O
R	Manufacture new product, first production run	P, Q
S	Conduct sales meetings	H, I
T	Contact customers and receive orders	O, S
U	Process orders and send the product	R, T

51. A company has decided to market a new product for the consumer market. The problems of how to plan and control the various phases of this project—sales promotion, training of salespeople, pricing, packaging, advertising, and manufacturing—are obvious to the management of this firm. They have asked you to guide then through this difficult venture using CPM, since time is of the essence. The first firm to market this type of product will reap substantial profits and will enhance its image by marketing such a revolutionary product. A list of the activities, with the expected time duration for each, is given in the table in terms of weeks.

Activity	*Description (Weeks)*	*Precedence*	*Time*
Manufacturing Activities			
A	Study equipment requirement	–	1
B	Select supplier of equipment	A	1
C	Study manufacturing procedures	B	3
D	Study quality control procedures	C	3
E	Study purchasing and inventory rules	B	3
F	Receive and install equipment	B	8
G	Place order for raw materials	E	1
H	Manufacture from raw materials for test and first production runs	G	4

I	Receive containers and packaging supplies	P	1
J	Have personnel available for first production run	F	0
K	Run manufacturing test	D, J, H, I, T	3
L	Run first production	K	1
Marketing Activities			
M	Price product	B, S	2
N	Do artwork for advertising	M	4
O	Send out advertising materials and packaging orders to suppliers	N	5
P	Produce advertising and packaging materials	O	1
Q	Hold sales meeting	K, S	2
R	Training salespeople	Q	3
Accounting Activities			
S	Determine cost of new product	B	1
T	Determine cost of the new product inventory	S	3

(a) Draw a network diagram for this project.

(b) Identify the critical path and determine its length.

52. A new type of water pump is to be designed for an automobile. Major specifications are given in the table. Draw the network diagram for the project.

Activity	*Description*	*Predecessor Activity*
A	Drawing prepared and approved	–
B	Cost analysis	A
C	Tool feasibility (economics)	A
D	Tool manufactured	C
E	Favourable cost	B, C
F	Raw materials procured	D, F
G	Sub-assemblies ordered	E
H	Sub-assemblies received	G
I	Parts manufactured	D, F
J	Final assembly	I, H
K	Testing and shipment	K

53. Listed in the table are the activities and sequencing required in the computerization of a bank branch.

Activity	*Description*	*Predecessor Activity*
A	Preparation/Deliberations in the department	–
B	Dialogue with the union	A
C	Discussion/Approval/Sanction of local management	A
D	Customer education	B
E	Preparing specifications for the system	C
F	Selection of staff	B
G	Order/Acquisition of system	D, E
H	Alterations in the branch premises	C
I	Training of staff	F
J	Wiring/Power-supply set-up of the branch	H
K	Transition	G, I, J
L	Parallel run	K

Draw the arrow diagram of activities of the above project.

54. The medical faculty of a university is considering to hold a faculty development programme. It has planned the following activities. Prepare a network diagram showing the inter-relationships of the various activities.

Activity	*Description*	*Predecessor Activity*
A	Design conference meetings and theme	–
B	Design front cover of the conference proceedings	A
C	Prepare brochure and send request for papers	A
D	Compile list of distinguished speakers/guests	A
E	Finalize brochure and print it	C, D
F	Make travel arrangements for distinguished speakers/guests	D
G	Despatch brochures	–
H	Receive papers for conference	G
I	Edit papers and assemble proceedings	F, H
J	Print proceedings	B, I

55. Activities and their description associated with the Haryali Watershed Project are listed below:

Activity	*Description*	*Predecessor Activity*
A	Consult General Water Board (GWB)	–
B	Motivate farmers	A
C	Identify beneficiaries spots	B
D	Obtain demand certificate from GWB	C
E	No objection certificate from GWB	C
F	Apply to DPAP	C
G	Forward to bank	D, E, F
H	Process application	G
I	Requisition drilling high	G
J	Release !oans	H
K	Apply for power-	H
L	Purchase pump sets	I, J
M	Drilling of walls	I, J
N	Construct storage tank	I, J
O	Prepare field channels	I, J
P	Test water for quality	M
Q	Install pump sets	L, P
R	Power connections	K
S	Tests	R

Draw the arrow diagram of activities of the above project.

56. Following table gives the list of various activities and their immediate predecessors involved in installation of CAT scanner in a hospital:

Activity	*Description*	*Immediate Predecessors*	*Expected Duration (days)*
A	Finalization of the layout plans	–	2
B	Demolition of the structures	A	6
C	Walls erection	B	12
D	Flooring	B	8
E	Electrical wiring	C	6

F.	Air conditioning dueling	C	4
G	Fire alarm installation	C	3
H	False ceiling and light fittings	E, F, G	10
I	Wall plastering and painting	H	9
J	Equipment installation	D, I	6
K	Calibration and testing	J	3
L	Final finishing	K	2
M	Handing over	L	1

(a) Prepare a network diagram for this project.

(b) Find the total time required to install CAT scanner.

(c) Calculate the total float for various non-critical activities.

57. A research and development department is developing a new power supply for a console television set. It has broken the job down into the following for:

Job	*Description*	*Immediate Predecessors*	*Expected Time (days)*
A	Determine output voltages	–	5
B	Determine whether to use solid state rectifiers	A	1
C	Choose rectifier	B	1
D	Choose filters	B	3
E	Choose transformer	C	1
F	Choose chassis	D	1
G	Choose rectifier mounting	C	1
H	Layout chassis	E, F	3
I	Build and test	G, H	10

(a) Draw a critical path scheduling arrow diagram, identifying jobs letters and associating times with each. Indicate the critical path.

(b) What is the minimum time for completion of the project?

58. The following maintenance job has to be performed periodically on the heat exchangers in a refinery:

Task	*Description (hours)*	*Precedence*	*Duration*
A	Dismantle pipe connections	–	14
B	Dismantle header, closure, and floating head front	A	22
C	Remove tube bundle	B	10
D	Clean bolts	B	16
E	Clean header and floating head front	B	12
F	Clean tube bundle	C	10
G	Clean shell	C	6
H	Replace tube bundle	F, G	8
I	Prepare shell pressure test	D, E, H	24
J	Prepare tube pressure test and make the final reassembly	I	16

(a) Draw an arrow diagram for this project.

(b) Identify the critical path. What is its length?

(c) Find the total float and free float for each task.

59. In Border Roads Organization, the vehicles and equipment are being overhauled in base workshop. The overhauling of vehicles involves the number of jobs as given in the following table.

Job	*Description Time*	*Immediate Predecessors*	*Expected (days)*
A	Dismantling of vehicle	–	3
B	Inspection and preparation of visual inspection on report	B	2
C	Stripping of minor assys	B	2
D	Stripping of major assys	B	3
E	Overhauling of engine	C	40
F	Cleaning of chassis	C, D	2
G	Cleaning of body and cabin	C. D	2
H	Repairing of gear box	C	2
I	Repairing of break system	D	1
J	Repairing of axle and propeller shaft	D	6

K	Repairing of road spring	D	2
L	Repairing of chassis	F	2
M	Repairing of body and cabin	G	5
N	Placing of chassis and fitment of engine	E, L	2
O	Fitment of major assys	N, H	1
P	Fitment of minor assys and cabin	N, I, J, K, M	2
Q	Checking and rectification	O, P	1
R	Mechanical inspection for passing	Q	1
S	Painting	R	1
T	Checking and rectification before final inspection	S	1
U	Final inspection and passing	T	1
V	Closing of job card	U	1

(a) Draw a network diagram for the overhauling project.

(b) Identify the critical path. What is its length?

(c) Find the total float and free float for each task.

60. The sales manager of Domestic Products Limited was informed by the R&D department about the completion of the prototype of a particular product. He consulted the production manager on the time taken to produce the first batch of the product, which is needed for demonstration in his sales promotion programme. He also decided to invite a few industrial representatives to the demonstration of this new product and through them to launch it in the market. The various activities involved in this marketing project, their descriptions, estimated duration (in days) and immediate predecessors are given in the following table:

Activity	*Description*	*Duration* (days)	*Immediate* Predecessors
A	Collect data on specifications and capabilities	4	–
B	Prepare operation manual	4	A
C	Chart out promotion programme	4	B
D	Make copies of manual and promotion material	9	B
E	Produce first batch for demonstration	16	B
F	Prepare list of press representatives	2	C

G	Chief executive's conference with Managers	1	C
H	Press representatives reach Bombay	2	F, G
I	Promotional meetings	4	D, H
J	Product demonstration	2	E, I
K	Press representatives return home	2	J

(a) Draw the network diagram for the given project.

(b) Identify the critical path. What is the maximum time required to complete the project.

(c) Find the total float and free float (if any) for all the non-critical activities.

61. The President of ABC Manufacturing Company has an opportunity to participate in a project that has a sales price of Rs. 90,000 but it must be completed within 8 weeks. This letter of intent was received Friday afternoon. Both the superintendent of production and the cost accountant came in on Saturday and completed the appropriate time and cost for you based upon past jobs. Since the President needs an answer at 8.30 a.m. on Monday (start of the 8th week), you have been requested to determine the profitability of the project on an 8 week basis. An answer at 8.30 a.m. Monday allows the firm to start the production order at 10.00 a.m. in order to stay within the 8 weeks demanded by the customer. The time and cost under normal conditions without crashing the project is based upon an 11 week basis. What answer should the president give the customer on Monday morning. A table of times and cost is given below:

62. Listed in the table are the activities and sequencing requirements necessary for the completion of a research report.

Activity	*Description*	*Precedence*	*Duration (weeks)*
A	Literature search	–	6
B	Formulation of hypothesis	5	
C	Preliminary feasibility study	B	2
D	Formal proposal	C	2
E	Field analysis	A, D	2
F	Progress report	D	1
G	Formal research	A, D	6

H	Data collection	E	5
I	Data analysis	G, H	6
J	Conclusions	1	2
K	Rough draft	G	4
L	Final copy	J, K	3
M	Preparation of oral presentation	L	1

(a) Draw a network diagram for this project.

(b) Find the critical path. What is its length?

(c) Find the total float and the free float for each non-critical activity.

63. Listed in the table are the activities and their simplified sequencing requirement for publishing a textbook.

Activity	*Description*	*Preceding Activity*	*Expected Completion Time (months)*
A	Write book	–	12
B	Design book	A	1
C	Edit manuscript	A	6
D	Checking editing	C	2
E	Accept design	B	1
F	Copy edit	D, E	2
G	Prepare artwork	D, E	4
H	Accept and correct artwork	G	1/2
I	Set galleys	F	4
J	Check and correct galleys	I	1
K	Full page proofs	H, J	2
L	Check and correct pages	K	1
M	Prepare index	K	1
N	Set and correct index	M	1/2
O	Check camera-ready copy	L, N	1/2
P	Print and bind book	0	1

(a) Construct the PERT network and find the critical path.

(b) For each non-critical activity, find total float and free float.

64. Patel Machinery Co. has been offered a contract to build and deliver nine extruding presses to the ABC Bottling Co. The contract price is contingent on meeting a specified delivery time, a bonus being given for early delivery. The marketing department has established the following cost and time information:

Activity	*Normal Time (weeks)*			*Normal Cost*	*Crash Time*	*Crash Cost*
	Optimistic	*Pressistic*	*Most Likely*	*(Rs.)*	*(weeks)*	*(Rs.)*
1-2	1	5	3	15,000	1	19,000
2-3	1	7	4	18,000	3	24,000
2-4	1	5	3	14,000	2	16,000
2-5	5	11	8	15,000	7	16,000
3-6	2	6	4	13,000	2	15,000
4-6	5	7	6	12,000	4	13,000
5-7	4	6	5	20,000	4	24,000
6-7	1	5	3	17,000	1	20,000

65. The normal delivery time is 16 weeks for a contract price of Rs. 1,24,000. Based on the probability for each of the following specified delivery time, recommend the delivery schedule that the Patel Machinery Co. should follow:

Contract Delivery Time (weeks)	*Contract Amount (Rs.)*
15	1,42,500
14	1,45,000
13	1,50,000
12	1,52,500

66. An electronics firm has signed a contract to install an instrument landing device at the local airport. The complete installation can be broken down into fourteen separate activities. Each activity (labelled A through N), its predecessor activities, normal time and cost and crash time and cost are given below. The contract specifies that the installation will be completed within 18 days. There is a penalty of Rs. 100 per day beyond the specified completion time.

Activity Cost	*Preceding*	*Normal Time (days)*	*Normal Cost (days)*	*Crash Time (Rs.)*	*Crash (Rs.)*
A	–	3	320	1	360
B	–	5	550	4	600
C	–	6	575	4	700
D	A	7	750	5	850
E	A	4	420	3	490
F	B, D	2	180	2	280
G	C	4	425	3	485
H	A	8	850	5	900
I	C	5	475	4	535
J	C	7	675	5	735
K	E, F, G	4	400	3	440
L	H, I	6	650	4	750
M	L	3	280	2	335
N	J, K	5	525	4	575

The project will require the positioning of one engineer till the completion of work. Monthly cost of each engineer for such job is Rs. 10,000. This will be considered as indirect cost.

67. American Bottle Company (ABC) produces several types of glass containers. They have recently reduced capacity at several of their plants.

    Glass manufacturing involves large, expensive machines (including ovens), several of which were turned off in the capacity reduction.

    The machines were hard to shut down and to start up. In the event of a surge in demand, they wanted to know how quickly they could start one.

    How quickly can they start a new oven using normal times? What is the fastest time in which a new oven can be started, and how much additional cost is involved?

Cost (Rs.) per Unit Predecessor Time (hour) Reduction	*Activity*	*Normal Time*	*Crash Time*	
–	A Preheat glass	8	8	C
–	B Preheat oven	12	12	D
400	C Obtain materials	4	2	–
200	D Check valves	4	2	–
200	E Check pressure seals	2	1	B
–	F Add glass to oven	2	2	A, E
500	G Prepare bottlemaker	6	3	E
	H Run test production	4	4	F, G
500	I Examine test quantity and make adjustments	4	2	H
	J Refill oven with glass	2	2	H

68. The time-cost estimates for the various activities of a project are given below:

Event	*Preceding Activities*	*Time (Weeks)* *Normal*	*Crash*	*Cost (Rs.)* *Normal*	*Crash*
A	–	8	6	8,000	10,000
B	–	7	5	6,000	8,400
C	A	5	4	7,000	8,500
D	B	4	3	3,000	3,800
E	A	3	2	2,000	2,600
F	D, E	5	3	5,000	6,600
G	C	4	3	6,000	7,000

The project manager wishes to complete the project in the minimum possible time. However, he is not authorized to spend more than Rs. 5,000 on crashing.

Suggest the least-cost schedule for achieving the objective of the project manager. Assume that there is no indirect or utility cost.

69. A company has recently won a contract for the installation of a die casting machine and its associated building construction work at a local factory of a large national firm of electronic engineers.

The following table gives the various activities involved in this job, their normal time and cost estimates and their crash time and cost estimates.

Activity	*Description*	*Predecessor*		*Normal*	*Crash*
		Time (days)	*Cost (Rs.)*	*Time (days)*	*Cost (Rs.)*
A	Prepare foundations and underground services and erect building frame structure	–	30	90.000	251,05,000
B	Fabricate part and assemble steel frames to support the machine	–	25	1,80,000	201,90,000
C	Collect die casting machine and its associated gear from the manufacturers	–	10	50,000	8 54,000
D	Assemble and check control gear	C	10	7,500	7 9,000
E	Fit control gear on steel frames and instal	B, D	10	4,200	10 4,200
F	Fit aluminium sheet wall claddings	A, E	20	20,000	16 30,000
G	Erect assembled plant on to prepared foundation and framework and connect services	A, E	35	28,000	30 35,500
H	Erect mechanical handling plant	B, D	20	12,000	18 15,000
I	Fit ventilation and fire protection system	F	20	14,000	15 24,000

If the variable overhead costs are Rs. 5,000 per day, determine the optimal project duration.

70. A small marketing project consists of the jobs in the table given below. With each job is listed its normal time and a minimum or

crash time (in days). The cost (in Rs. per day) of crashing each job is also given.

Job	*Normal Duration (days)*	*Crash Duration (days)*	*Cost of Crashing (Rs. per Day)*
1-2	9	6	20
1-3	8	5	25
1-4	15	10	30
2-4	5	3	10
3-4	10	6	15
4-5	2	1	40

(a) What is the normal project length and the minimum project length?

(b) Determine the minimum crashing costs of schedules ranging from normal length down to, and including the minimum length schedule, *i.e.*, if L is the length of the normal schedule, find the costs of schedules which are L, L – 1, L – 1 days long and so on.

Overhead cost for the project is Rs. 60 per day. What is the optimal length schedule duration of each job for your solution?

71. Following table gives the list of various activities involved in the production of a wireless communication equipment, their immediate predecessor(s), their normal time and cost estimates and their crash time and cost estimates:

Activity	*Description*	*Time (months)*		*Cost (Rs.)*		*Predecessor*
		Normal	*Crash*	*Normal*	*Crash*	
A	System calculations	3	1	45,000	63,000	–
B	Release of drawings	2	1	15,000	23,000	–
C	Procurement of PSU	12	10	85,000	1,05,400	A, B
D	Procurement of raw material	8	6	4,50,000	5,00,000	A, B
E	Production documentation	4	3	40,000	49,500	A, B
F	Time study/shop order	3	1	25,000	41,000	E
G	PCB Manufacture	6	5	50,000	59,000	D, F

H	Mechanical parts manufacturers	7	6	2,00,000	2,20,000	D, E, F
I	Electronic assembly	2	2	43,000	43,000	C, G, H
J	Testing	4	2	50,500	75,000	1

Indirect costs per month are Rs. 20,000. Determine the optimal time versus cost schedule.

72. The Sales Manager of Domestic Products Limited, Bombay was informed by the R&D department about the completion of the prototype of a particular product. He consulted the production manager on the time taken to produce the first batch of the product, which needed for demonstration in his sales promotion programme. He also decided to invite a few industrial representatives to the demonstration of this new product and through them to launch it in the market. The various activities involved in this marketing project, their descriptions, estimated durations (in days) and immediate predecessors are given in the following tables:

Activity	*Description*	*Duration (days)*	*Predecessor*
A	Collect data on specifications and capabilities	4	–
B	Prepare operation manual	4	A
C	Chart out promotion programme	4	B
D	Make copies of manual and promotion material	9	B
E	Produce first batch for demonstration	16	B
F	Prepare list of press representatives	2	C
G	Chief executives conference with managers	1	C
H	Press representatives reach Bombay	2	F, G
I	Promotional meetings	4	D, H
J	Product demonstration	2	E, I
K	Press representatives return home	2	J

(a) Determine the maximum time required to complete the above project, list the critical activities and find the total float and free float, if any, for all the non-critical activities.

(b) If the indirect cost per day for the project under consideration is Rs. 300, the normal and crash time and cost estimate for various activities are as given in the following table. Determine the optimal project duration.

Activity	*Normal*		*Crash*	
	Time (days)	*Cost (Rs.)*	*Time (days)*	*Cost (Rs.)*
A	4	100	3	450
B	4	150	2	510
C	4	200	4	200
D	9	500	4	1,000
E	16	2,000	8	2,960
F	2	60	1	140
G	1	100	1	100
H	2	2,500	1	6,000
I	4	2,200	3	2,340
J	2	700	2	700
K	2	2,500	1	6,000

73. A project has the following activities and their durations.

Activity :	12	13	14	24	25	34	36	47	57	67	68	78
Duration (days):	13	15	9	10	27	7	18	30	12	10	10	9

(a) Draw the network of the project and find its duration.

(b) At the end of 25 days it is observed that

(i) activities 12, 13, 14 have been completed.

(ii) activity 24 is being done and will be completed in 5 more days.

(iii) activity 36 is in progress and will need 20 more days for completion.

(iv) activity 67 is presenting some problem and will take 15 days.